AF568063

GLOBAL ENVIRONMENT

Issues and Problems

GLOBAL ENVIRONMENT

Issues and Problems

Edited by
K.S. Sengar
G.S. Chauhan
Subhash Chand
Ramesh Kumar

L.G. PUBLISHERS DISTRIBUTORS

First Published, 2012

ISBN 978-81-910382-1-7

Published by
L.G. PUBLISHERS DISTRIBUTORS
49, Gali No. 14, Pratap Nagar
Mayur Vihar Phase I, Delhi 110 091

Printed at
Arpitprintographers, Delhi 110 032
arpitprinto@yahoo.com

Contents

Introduction

The sum of all physical, chemical, biological and social factors which compose the surroundings of man is referred to as environment and each element of these surroundings constitutes a resource on which man draws in order to develop a better life. Thus any part of our natural environment such as land, water, air, minerals, forest, rangeland, wildlife or even human population can be used to promote his welfare for socio-economic and cultural needs, both at the individual and community level. But man ruthlessly exploited his own surroundings to meet his day-to-day requirements. The story still continues in various parts of the world at a slower or faster pace.

Perhaps the most amazing feature of our planet is the rich diversity of life that exists here. Millions of beautiful and intriguing species populate the earth and help to sustain a habitable environment. The vast magnitude of life creates complex, interrelated communities, where towering trees and huge animals live together with, and depend upon, micro life forms such as viruses, bacteria and fungi. Together, all these organisms make up delightfully diverse self-sustaining communities, including dense, moist forests, vast sunny savannas and richly coral reefs.

As human population grew, it lived unwittingly at the cost of other living things and ecosystems. They did what all animals do naturally to survive where survival normally sets the parameters for checks and balance on living things. This changed with humans. Human beings are adaptable, and they have no natural predators.

The result had been a gradual and healthy population growth till it exploded in recent history. The more people there are, the more vehicles, houses, schools, food and space are demanded. With this explosion, human needs and demands of earth's resources increased exponentially, exhausting many resources, causing the extinction of other species, and creating conditions that have become unhealthy for even humans.

Population has grown at an alarming rate in the past century. In the decades, most of the growth will be in developing countries where resources and services are already strained by present populations. Whether there are sufficient resources to support such populations on a sustainable basis is one of the most important questions we face. How we might stabilize population and what level of resources consumption we and future generations can afford are an equally difficult part of this challenging equation.

Food shortages in many places may increase in frequency and severity if population growth, soil erosion and nutrients depletion continue at the same rate in the future as they have in the past. Water deficits and contamination of existing water supplies threaten to be critical environmental issues in the future for agricultural production as well as domestic and industrial uses. Many countries already have serious water shortages and more than one billion people lack access to clean water or adequate sanitation. Violent conflicts over control of natural resources may crop up in many places if we do not learn to live within nature's budget.

How we obtain and use energy is likely to play a crucial role in our environmental future. Fossil fuels (oil, coal and natural gas) presently supply about 60% of energy used in industrialized countries. Increased population and over use are responsible for depletion of fossil fuel the world over. Supplies of these fuels are diminishing at an alarming rate and problems associated with their acquisition and use—air, water and soil pollution, mining damage, shipping accidents, and political insecurity may limit where and how we use the remaining resources. Cleaner renewable energy resources: solar power, wind and biomass together with conservation, may replace environmentally destructive energy resources if we invest in appropriate technology in the next few years.

Use of fossil fuels in vehicles, industries and power generation plants release carbon dioxide and other heat absorbing gases that cause global warming and may bring about sea level rises and catastrophic climate changes. Acid formed in the air as a result of fossil fuel combustion have already caused extensive damage to building materials, plants as well as animals life in sensitive ecosystems in many places. Continued fossil fuel use without pollution control measures could cause even more extensive damage. Chlorinated compounds such as chlorofluorocarbons used in refrigeration and air conditioning, also contribute to global warming, as well as damaging the stratospheric ozone layer that protects us from cancer causing ultraviolet radiation in sunlight.

Toxic air, water and soil pollutants along with mountains of solid and hazardous wastes are becoming overwhelming problems in industrialized countries. We produced hundreds of millions of tons of these dangerous materials annually, and noxious stuff dumped in their own backyard, but too often the solution is to export it to someone else's. We may come to a political impasse, where our failure in industries result in wastes. How to dispose of them safely will stop pollution, toxic wastes, stress and other environmental ills of modern society have thus become a greater threat than infectious diseases for many of us in industrialized countries.

Psychologists have now realized that all of us greatly need contact with the natural world. We live in crowded places. Dirtiness and disorderliness would affect the mental as well as physical sickness, because it is undetectable and untreatable. Thus open green spaces are vital in cities as "lungs" have become more essential than ever. They reduce the dirt and pollution of the city and improve the health standards of its inhabitants.

Biological diversity is threatened by various forms of pollution but ultimately the greatest threat may be global climate change. Species and ecosystems within protected areas will certainly be affected by climate change. Destruction of tropical forests, coral reefs, wetlands and other biologically rich landscapes is causing an alarming loss of species and reduction of biological variety and abundance that could severely limit our future of options.

Many rare and endangered species are threatened directly or indirectly by human activities. In addition to practical values, and habitat necessary for their survival, forest is the natural cover for our land. A good forest cover implies that all is well with the land, all the natural forces—water, air, soil and animal life are interacting to produce the maximum amount of living material in terms of biomass. The soil is kept healthy and vigorous by leaf litter and other decaying vegetation. The self-renewing cycle of seeding, growth, maturity, decay, death is perpetual and a forest, if we do not interfere with it too much can be ever lasting. People live in great poverty and forests are an open treasure-house. People feel that it is their right to walk into collect fire wood, to graze their cattle and to take whatever they fancy from the forest. In any country tree and wood stealing is a sign that all is not well with a nation's economic life. The biggest causes of forest denudation are of course clearing of forest for deforestation, agricultural activities, fuel gathering, cattle grazing and very often paper is made from bamboo, and our large paper mills are using up more bamboos than forest can replace.

A new threat in the third world to forest health is from the invasion of imported plants which do not rightfully belong to a country. Once an exotic species has captured the forest, it is virtually impossible to free it.

Gene pollution is also especially problematic for the southern countries where the centre of origin is for many crops. In these areas traditional crop varieties could become polluted with genes from the genetically engineered crops. If genes flow into population of wild relatives that enhance their fitness, super weeds could be created. The BT crops could upset established ecological balances by either causing the wild plants to flourish excessively and become a weed, or be reducing the insect population that previously fed on the newly toxic plants.

Hence we would like to suggest to all researchers, pioneer scientists and professors that the globe is in our hands and remedies have to be made before nature alarms us finally.

1

Environmental Issues in India

Niharika Sengar

The rapid growing population and *economic development* are leading to the *environmental degradation in India because* of the uncontrolled growth of *urbanization* and *industrialization*, expansion and massive intensification of *agriculture*, and the destruction of forests.

Major environmental issues are Forest and Agricultural Land Degradation; Resource Depletion (water, mineral, forest, sand, rocks etc.,), Environmental Degradation, Public Health, Loss of Biodiversity, Loss of Resilience in Ecosystems, Livelihood Security for the Poor.

It is estimated that the country's population will increase to about 1.26 billion by the year 2016. The projected population indicates that India will be the first most populous country in the world and China will be ranking second in the year 2050. India having 18% of the world's population on 2.4% of the world's total area has greatly increased the pressure on its natural resources. Water shortages, soil exhaustion and erosion, deforestation, air and water pollution afflicts many areas.

India's *water supply and sanitation issues* are related to many environmental issues.

Major Issues

One of the primary causes of environmental degradation in a country could be attributed to rapid growth of population, which adversely affects the natural resources and environment. The rising population and the environmental deterioration face the challenge of sustainable development. The existence or the absence of favourable natural resources can facilitate or retard the process of socio-economic development. The three basic demographic factors of births (natality), deaths (mortality) and human migration (migration) and immigration (population moving into a country produces higher population) produce changes in population size, composition, distribution and these changes raise a number of important questions of cause and effect.

Population growth and economic development are contributing to many serious environmental calamities in India. These include heavy pressure on land, land degradation, forests, habitat destruction and loss of bio-diversity. Changing consumption pattern has led to rising demand for energy. The final outcomes of this are air pollution, global warming, climate change, water scarcity and water pollution.

Environmental issues in India include various natural hazards, particularly cyclones and annual monsoon floods, population growth, increasing individual consumption, industrialization, infrastructural development, poor agricultural practices, and resource maldistribution have led to substantial human transformation of India's natural environment. An estimated 60% of cultivated land suffers from soil erosion, waterlogging, and salinity. It is also estimated that between 4.7 and 12 billion tons of topsoil are lost annually from soil erosion. From 1947 to 2002, the average annual per capita water availability declined by almost 70% to 1,822 cubic metres, and overexploitation of groundwater is problematic in the states of Haryana, Punjab, and Uttar Pradesh. Forest area covers 18.34% of India's geographic area (637000 km^2). Nearly half the country's forest cover is found in the state of Madhya Pradesh (20.7%) and the seven states of the northeast (25.7%); the latter is experiencing net forest loss. Forest cover is declining because of harvesting for fuel wood and the expansion of agricultural land.

These trends, combined with increasing industrial and motor vehicle pollution output, have led to atmospheric temperature increases, shifting precipitation patterns, and declining intervals of drought recurrence in many areas.

The Indian Agricultural Research Institute of Parvati has estimated that a 3°C rise in temperature will result in a 15 to 20% loss in annual wheat yields. These are substantial problems for a nation with such a large population depending on the productivity of primary resources and whose economic growth relies heavily on industrial growth. Civil conflicts involving natural resources most notably forests and arable land have occurred in eastern and northeastern states.

Water Pollution

Main article: *Water Supply and Sanitation in India*

Out of India's 3,119 towns and cities, just 209 have partial treatment facilities; and only 8 have full wastewater treatment facilities (WHO 1992). About 114 cities dump untreated sewage and partially cremated bodies directly into the *Ganges River.* Downstream, the untreated water is used for drinking, bathing, and washing. This situation is typical of many rivers in India as well as other developing countries.

Open *defecation* is widespread even in urban areas of India.

Water resources have not therefore been linked to either domestic or international violent conflict as was previously anticipated by some observers. Possible exceptions include some communal violence related to distribution of water from the *Kaveri River* and political tensions surrounding actual and potential population displacements by dam projects, particularly on the *Narmada River.*

Ganges

Millions depend on the polluted *Ganges river.*

Main article: *Pollution of Ganga.*

To know why 1,000 Indian children die of diarrhoeal sickness every day, take a wary stroll along the Ganges in *Varanasi.* As it enters the city, Hinduism's sacred river contains 60,000 faecal coliform bacteria per 100 millilitres, 120 times more than is considered safe for bathing. Four miles downstream, with inputs from 24 gushing sewers and 60,000 pilgrim bathers, the concentration is 3,000 times over the safety limit. In places, the Ganges becomes black and septic. Corpses, of semi-cremated adults or enshrouded babies, drift slowly by.

— *The Economist* on December 11, 2008

More than 400 million people live along the *Ganges* River. An estimated 2,000,000 persons ritually bathe daily in the river, which is considered holy by Hindus. In the Hindu religion it is said to flow from the lotus feet of *Vishnu* (for *Vaisnava* devotees) or the hair of *Shiva* (for *Saivites*). The spiritual and religious significance could be compared to what the Nile river meant to the ancient Egyptians. While the Ganges may be considered holy, there are some problems associated with the ecology. It is filled with chemical wastes, sewage and even the remains of human and animal corpses which carry major health risks by either direct bathing in the water (e.g.: *Bilharziasis*_infection), or by drinking (the *faecal oral route*).

Yamuna

Main article: *Yamuna*

News Week describes Delhi's sacred *Yamuna River* as "a putrid ribbon of black sludge" where faecal bacteria is 10,000 over safety limits despite a 15-year programme to address the problem. *Cholera* epidemics are not unknown.

Air Pollution

See also: *Global warming in India*

Indian cities are polluted by vehicles and industry emissions. Road dust due to vehicles also contribute up to 33% of *air pollution* In cities like *Bangalore*, around 50% of children suffer from *asthma*. India has an *emission standard* of Bharat Stage II (*Euro II*) for vehicles since 2005.

One of the biggest causes of air pollution in India is from the transport system. Hundreds of millions of old diesel engines continuously burning away diesel which has anything between 150 to 190 times the amount of sulphur that European diesel has. Of course the biggest problems are in the big cities where there are huge concentrations of these vehicles. On the positive side, the government appears to have noticed this massive problem and the associated health risks for its people and is slowly but surely taking steps. The first of which was in 2001 when it ruled that its entire public transport system, excluding the trains, be converted from diesel to compressed gas (CPG). Electric rickshaws are being designed and will be subsidized by the government but the supposed ban on the cycle rickshaws in Delhi will require a huge increase on the reliance of other methods of transport, mainly those with engines.

Another major cause of air pollution is due to cremations in India. In India 78% of the population consign the dead bodies to fire for cremation as a ritual in open air. Traditionally they have been using butter or ghee and a few herbs while the body is confined to fire. These are required since the wood fire temperature does not go beyond 300 C or 600 F, but when the butter or ghee is added the temperature obtained is upto 700 C or 1400 F, which has been proved now scientifically to be the optimum temperature required for cremation of a human body. Just as the low temperature creates pollution; higher temperature is also found to create pollution with emissions dangerously harmful for the environment.

By consigning the corpse to fire, these pollution risks are reduced and if, in that fire some ghee and havan samagri are added, the practice and experiments have established that there is less environmental pollution and emission of foul smell because of their disinfecting properties. By adding ghee to the fire, the rise in temperature of the flames results in total destruction of those germs and worms.

Paryavaran Sanrakshan Nyas, a non-government voluntary organization of Chandigarh (India), chose to undertake this task which had escaped the attention of the people in the urbanized cities. In rural areas in villages even today, they use a lot of ghee, herbs and cow dung (which is a strong anti-pollution agent when burnt) to arrest this pollution. Besides, the cremation grounds in the villages are placed at far isolated areas, away from the populated localities. In cities, the situation is different. The cremation grounds are mostly located in and around the inhabited areas affecting seriously the living population. (*Pollution through Cremation* by Savita Sethi published by Paryavaran Sanrakshan Nyas 2005)

Aware of all these factors and the problems, the four women Savita Sethi, Sudesh Gupta, Prem Lata Duggal and Usha Ghai of Chandigarh thought of the issue and decided to fight this un-noticed pollution being caused in the 'City Beautiful' and create awareness amongst the residents. To carry out the mission they decided to form a Trust and elicit support and cooperation from elite and awakened members of the society. Subsequently a Trust under the name of Paryavaran Sanrakshan Nyas was registered at Chandigarh with nine Trustees of the Nyas. (*Pollution through Cremation* by Savita Sethi published by Paryavaran Sanrakshan Nyas 2005)

The Trust believed that besides contributing to this noble social cause of pollution control, a respectful and appropriate adieu could be also given, to the departed soul of those unprivileged people who are not able to bear this bare minimum for the last rites of their beloved ones. The Trustees decided that on every cremation the Trust should contribute one kg. of pure ghee and five kgs. of havan samagri (a mixture of organic herbs having ingredients which have anti-pollutant, disinfectant, aromatic, nourishing and nutritive qualities) a voluntary contribution of 5 kgs. of havan samagri mixed in 1 kg. of desi ghee on every cremation of any caste, creed or faith at the Chandigarh Crematorium and thus save the city from such threatened possible pollution.

It also appeared that the excessive pollution was having an adverse effect on the Taj Mahal. After a court ruling all transport in the area was shut down shortly followed by the closure of all industrial factories in the area. The air pollution in the big cities is rising to such an extent that it is now 2.3 higher than the amount recommended by WHO (World Health Organization). *(Pollution through Cremation* by Savita Sethi published by Paryavaran Sanrakshan Nyas 2005)

Noise Pollution

The Supreme Court of India gave a significant verdict on *noise* pollution in 2005. Unnecessary honking of vehicles makes for a high decibel level of noise in cities. The use of loudspeakers for political purposes and by temples and mosques make for noise pollution in residential areas.

2

A Study on the Rapidly Growing Environmental Problems of India

Tanu Middha, Sanjeev Gupta and Keshav Singh Gurjar

1. Introduction

Electronic waste, popularly known as 'e-waste' can be defined as electronic equipment/products connected with power plug batteries which have become obsolete due to:

1) Advancement in technology,
2) Changes in fashion, style and status,
3) Nearing the end of their useful life.

The electronic industry is the world's largest and fastest growing manufacturing industry (Radha, 2002; DIT, 2003). During the last decade, it has assumed the role of providing a forceful leverage to the socio-economic and technological growth of a developing society. The consequence of its consumer-oriented growth combined with rapid product obsolescence and technological advances are a new environmental challenge the growing menace of "Electronics Waste" or "e-waste" that consists of obsolete electronic devices. It is an emerging problem as well as a business opportunity of increasing significance, given the volumes of e-waste being generated and the content of both toxic and valuable materials in them. The fraction including iron, copper, aluminium, gold and other metals in e-waste is over 60%, while plastics account for about 30% and the hazardous pollutants comprise only about 2.70% (Widmer et al., 2005). Solid waste management, which is already a mammoth task in India, is becoming more complicated by the invasion of e-waste, particularly computer waste. E-waste from developed countries find an easy way into developing countries in the name of free trade (Toxics Link, 2004) is further complicating the problems associated with waste management. The paper highlights the associated issues and strategies to address this emerging problem, in the light of initiatives in India.

Electronic waste (e-waste) comprises waste electronic goods which are not fit for their originally intended use such as air conditioners, cellular phones, personal stereos, computers. E-waste contains toxic substances and chemicals, which are likely to have an adverse effect on environment and health. If not handled properly e-waste is hazardous only if it contains hazardous constituents.

Electronic waste, e-waste, e-scrap, or Waste Electrical and Electronic Equipment (WEEE) describes loosely discarded, surplus, obsolete, broken, electrical or electronic devices. The processing of electronic waste in developing countries causes serious health and pollution problems because electronic equipment contains some very serious contaminants such as *lead, cadmium, beryllium* and *brominates flame retardants*. Even in developed countries recycling and disposal of e-waste involves significant risk to workers and communities and great care must be taken to avoid unsafe exposure in recycling operations and leaching of material such as heavy metals from *landfills* and *incinerator* ashes. "E-waste" is a popular, informal name for electronic products nearing the end of their useful life." E-wastes are considered dangerous, as certain components of some electronic products contain materials that are hazardous, depending on their condition and density. The hazardous content of these materials pose a threat to human health and environment. Discarded computers, televisions, VCRs. stereos, copiers, fax machines, electric lamps, cell phones, audio equipment and batteries if improperly disposed can lead to other substances contaminating soil and groundwater. Many of these products can be reused, refurbished, or recycled in an environmentally sound manner so that they are less harmful to the ecosystem.

Definition of E-waste

Electronic waste, popularly known as 'e-waste' can be defined as electronic equipment/products connected with power plug batteries which have become obsolete due to: advancement in technology changes in fashion, style and status hearing the end of their useful life.

Classification of E-waste

E-waste encompasses an ever growing range of obsolete electronic devices such as computers, servers, main frames, monitors, TVs and display devices, telecommunication devices such as cellular phones and pagers, calculators, audio and video devices, printers, scanners, copiers and fax machines besides refrigerators, air conditioners, washing converters, electronic components such as chips, processors, mother boards, printed machines, and microwave ovens, e-waste also covers recording devices such as DVDs, CDs, floppies, tapes, printing cartridges, military electronic waste, automobile catalytic circuit boards, industrial electronics such as sensors, alarms, sirens, security devices and automobile electronic devices.

2. Indian Scenario

There is an estimate that the total obsolete computers originating from government offices, business houses, industries and household is of the order of 2 million tons. Manufacturers and assemblers in a single calendar year are estimated to produce around 1200 tons of electronic scrap. It should be noted that the obsolence rate of personal computers(PC) is one in every two years. The consumers find it convenient to buy a new computer rather than upgrade the old one due to the changing configuration, technology and the attractive offers of the manufacturers. Due to and the lack of governmental legislations on e-waste, standards for disposal, proper mechanism for handling these toxic hi-tech products, mostly end up in landfills or partly recycled in unhygienic conditions and partly thrown into waste streams. Computer waste is generated from the individual households; the government, public and private sectors; computer retailers; manufacturers; foreign embassies; secondary markets of old PCs. Of these, the biggest source of PC scrap are foreign countries that export huge computer waste in the form of reusable components. Electronic waste or e-waste is one of the rapidly growing environmental problems of the world. In India, the electronic waste management assumes greater significance not only due to the generation of our own waste but also dumping of e-waste particularly computer waste from the developed countries.

With extensively using computers and electronic equipment and people dumping old electronic goods for new ones, the amount of e-waste generated has been steadily increasing. At present, Bangalore alone generates about 8000 tons of computer waste annually and in the absence of proper disposal, they find their way to scrap dealers.

E-Parisaraa, an eco-friendly recycling unit on the outskirts of Bangalore which is located in Dobaspet industrial area, about 45 kms north of Bangalore, makes full use of e-waste. The plant which is India's first scientific e-waste recycling unit will reduce pollution, landfill waste and recover valuable metals, plastics and glass from waste in an eco-friendly manner. E-Parisaraa has developed a circuit to extend the life of tube lights. The circuit helps to extend the life of fluorescent tubes by more than 2000 hours. If the circuits are used, tube lights can work on lower voltages: The initiative is to aim at reducing the accumulation of used and discarded electronic and electrical equipment.

India as a developing country needs simpler, low cost technology keeping in view maximum resource recovery in environmental friendly methodologies. E-Parisaraa, deals with the practical aspect of e-waste processing as mentioned below by hand. Phosphor affects the display resolution and luminance of the images that is seen in the monitor.

E-Parisaraa's Director P. Parthasarathy, an IIT Madras graduate; and a former consultant for a similar e-waste recycling unit in

Singapore, has developed an eco-friendly methodology for reusing, recycling and recovery of metals, glass and plastics with non-incineration methods. The hazardous materials are segregated separately and sent for secure land fill for ex.: phosphor coating, LEDs, mercury, etc. We have the technology to recycle most of the e-waste and only less than one per cent of this will be regarded as waste, which can go into secure landfill planned in the vicinity by the HAWA project.

3. Research Methodology

Quantitative Research: Quantitative research is used to measure how many people feel, think or act in a particular way. These surveys tend to include large samples, anything from 50 to any number of interviews. In *quantitative research* your aim is to determine the relationship between one thing (an independent variable) and another (a dependent or outcome variable)

Secondary information is taken through different research articles, books, journals published articles of different eminent professors and readers.

4. Effects on Environment and Human Health

Disposal of e-wastes is a particular problem faced in many regions across the globe. Computer wastes that are landfilled produces contaminated leachates which eventually pollute the groundwater. Acids and sludge obtained from melting computer chips, if disposed on the ground causes acidification of soil. For example, Guiyu, Hong Kong, a thriving area of illegal e-waste recycling is facing acute water shortages due to the contamination of water resources.

This is due to disposal of recycling wastes such as acids, sludges, etc. in rivers. Now water is being transported from faraway towns to cater to the demands of the population. Incineration of e-wastes can emit toxic fumes and gases, thereby polluting the surrounding air. Improperly monitored landfills can cause environmental hazards. Mercury will leach when certain electronic devices, such as circuit breakers are destroyed. The same is true for polychlorinated biphenyls (PCBs) from condensers. When brominated flame retardant plastic or cadmium containing, plastics are landfilled, both polybrominated dlphenyl ethers (PBDE) and cadmium may leach into the soil and groundwater. It has been found that significant amounts of lead ion are dissolved from broken lead containing glass, such as the cone glass of cathode ray tubes, gets mixed with acid waters and are a common occurrence in landfills.

Not only does the leaching of mercury pose specific problems, the vaporization of metallic mercury and dimethylene mercury, both part of Waste Electrical and Electronic Equipment (WEEE) is also of concern. In addition, uncontrolled fires may arise at landfills and this could be a frequent occurrence in many countries. When exposed to fire, metals and other chemical substances, such as the extremely toxic dioxins and furans (TCDD tetrachloro dibenzo-dioxin, PCDDs-polychlorinated dibenzodioxins. PBDDs- polybrominated dibenzo dioxin and PCDFspoly chlorinated dibenzo furans) from halogenated flame retardant products and PCB containing condensers can be emitted. The most dangerous form of burning e-waste is the open air burning of plastics in order to recover copper and other metals. The toxic fall out from open air burning affects both the local environment and broader, global air currents, depositing highly toxic by products in many places throughout the world.

If these electronic items are discarded with other household garbage, the toxics pose a threat to both health and vital components of the ecosystem. In view of the ill effects of hazardous wastes to both environment and health, several countries exhorted the need for a global agreement to address the problems and challenges posed by hazardous waste. Also, in the late 1980s, a tightening of environmental regulations in industrialized countries led to a dramatic rise in the cost of hazardous waste disposal. Searching for cheaper ways to get rid of the wastes, "toxic traders" began shipping hazardous waste to developing countries. International outrage following these irresponsible activities led to the drafting and adoption of strategic plans and regulations at the Basel Convention. The Convention secretariat, in Geneva, Switzerland, facilitates the implementation of the Convention and related agreements. It also provides assistance and guidelines on legal and technical issues, gathers statistical data, and conducts training on the proper management of hazardous waste.

5. Management of E-wastes

It is estimated that 75% of electronic items are stored due to uncertainty of how to manage it. These electronic junks lie unattended in houses, offices, warehouses, etc. and normally mixed with household wastes, which are finally disposed of at landfills. This necessitates implementable management measures.

In industries management of e-waste should begin at the point of generation. This can be done by waste minimization techniques and by sustainable product design. Waste minimization in industries involves adopting:

- inventory management,
- production process modification,
- volume reduction,
- recovery and reuse.

(I) Inventory Management

Proper control over the materials used in the manufacturing process is an important way to reduce waste generation (Freeman, 1989). By reducing both the quantity of hazardous materials used in the process and the amount of excess raw materials in stock, the quantity of waste generated can be reduced. This can be done in two ways, i.e. establishing material purchase review and control procedures and inventory tracking system.

Developing review procedures for all material purchased is the first step in establishing an inventory management programme. Procedures should require that all materials be approved prior to purchase. In the approval process all production materials are evaluated to examine if they contain hazardous constituents and whether alternative non-hazardous materials are available.

Another inventory management procedure for waste reduction is to ensure that only the needed quantity of a material is ordered. This will require the establishment of a strict inventory tracking system. Purchase procedures must be implemented which ensure that materials are ordered only on an as needed basis and that only the amount needed for a specific period of time is ordered.

(II) Production Process Modification

Changes can be made in the production process, which will reduce waste generation. This reduction can be accomplished by changing the materials used to make the product or by the more efficient use of input materials in the production process or both. Potential waste minimization techniques can be broken down into three categories:

i) Improved operating and maintenance, procedures,

ii) Material change and

iii) Process equipment modification.

Improvements in the operation and maintenance of process equipment can result in significant waste reduction. This can be accomplished by reviewing current operational procedures or lack of procedures and examination of the production process for ways to improve its efficiency. Instituting standard operation procedures can optimize the use of raw materials in the production process and reduce the potential for materials to be lost through leaks and spills. A strict maintenance programme, which stresses corrective maintenance, can reduce waste generation caused by equipment failure. An employee-training programme is a key element of any waste reduction programme. Training should include correct operating and handling procedures, proper equipment use, recommended maintenance and inspection schedules, correct process control specifications and proper management of waste materials.

Hazardous materials used in either a product formulation or a production process may be replaced with a less hazardous or non-hazardous material. This is a very widely used technique and is applicable to most manufacturing processes. Implementation of this waste reduction technique may require only some minor process adjustments or it may require extensive new process equipment. For example, a circuit board manufacturer can replace solvent based product with water based flux and simultaneously replace solvent vapour degreaser with detergent parts washer.

Installing more efficient process equipment or modifying existing equipment to take advantage of better production techniques can significantly reduce waste generation. New or updated equipment can use process materials more efficiently producing less waste. Additionally such efficiency reduces the number of rejected or off specification products, thereby reducing the amount of material which has to be reworked or disposed of. Modifying existing process equipment can be a very cost effective method of reducing waste generation. In many cases the modification can just be relatively simple changes in the way the materials are handled within the process to ensure that they are not wasted. For example, in many electronic manufacturing operations, which involve coating a product, such as electroplating or painting, chemicals are used to strip off coating from rejected products so that they can be recoated. These chemicals, which can include acids, caustics, cyanides, etc. are often a hazardous waste and must be properly managed. By reducing the number of parts that have to be reworked, the quantity of waste can be significantly reduced.

(III) Volume Reduction

Volume reduction includes those techniques that remove the hazardous portion of a waste from a non-hazardous portion. These techniques are usually to reduce the volume, and thus the cost of disposing of a waste material. The techniques that can be used to reduce waste stream volume can be divided into two general categories: source segregation and waste concentration. Segregation of wastes is in many cases a simple and economical technique for waste reduction. Wastes containing different types of metals can be treated separately so that the metal value in the sludge can be recovered. Concentration of a waste stream may increase the likelihood that the material can be recycled or reused.

Methods include gravity and vacuum filtration, ultra filtration, reverse osmosis, freeze vaporization, etc.

For example, an electronic component manufacturer can use compaction equipment to reduce volume of waste cathode ray tube.

(IV) Recovery and Reuse

This technique could eliminate waste disposal costs, reduce raw material costs and provide income from a salable waste. Waste can be recovered on site, or at an off site recovery facility, or through inter industry exchange. A number of physical and chemical techniques are available to reclaim a waste material such as reverse osmosis, electrolysis, condensation, electrolytic recovery, filtration, centrifugation, etc. For example, a printed-circuit board manufacturer can use electrolytic recovery to reclaim metals from copper and tin lead plating bath.

However recycling of hazardous products has little environmental benefit if it simply moves the hazards into secondary products that eventually have to be disposed of. Unless the goal is to redesign the product to use non-hazardous materials, such recycling is a false solution.

(V) Sustainable Product Design

Minimization of hazardous wastes should be at product design stage itself keeping in mind the following factors :

- ***Rethink the product design:*** Efforts should be made to design a product with fewer amounts of hazardous materials. For example, the efforts to reduce material use are reflected in some new computer designs that are flatter, lighter and more integrated. Other companies propose centralized networks similar to the telephone system.
- ***Use of renewable materials and energy:*** Bio based plastics are plastics made with plant-based chemicals or plant-produced polymers rather than from petro-chemicals. Bio-based toners, glues and inks are used more frequently. Solar computers also exist but they are currently very expensive.
- ***Use of non-renewable materials that are safer:*** Since many of the materials used are non-renewable, designers could ensure the product is built for reuse, repair and/or upgradeability. Some computer manufacturers such as Dell and Gateway lease out their products thereby ensuring they get them back to further upgrade and lease out again.

6. Management Option

Considering the severity of the problem, it is imperative that certain management options be adopted to handle the bulk e-wastes. The following are some of the management options suggested for the government, industries and the public.

7. Suggestions

(I) Responsibilities of the Government

(*i*) Governments should set up regulatory agencies in each district, which are vested with the responsibility of coordinating and consolidating the regulatory functions of the various government authorities regarding hazardous substances.

(ii) Governments should be responsible for providing an adequate system of laws, controls and administrative procedures for hazardous waste management (Third World Network: 1991). Existing laws concerning e-waste disposal be reviewed and revamped. A comprehensive law that provides e-waste regulation and management and proper disposal of hazardous wastes is required. Such a law should empower the agency to control, supervise and regulate the relevant activities of government departments.

Under this law, the agency concerned should

- Collect basic information on the materials from manufacturers, processors and importers and to maintain an inventory of these materials. The information should include toxicity and potential harmful effects.
- Identify potentially harmful substances and require the industry to test them for adverse health and environmental effects.
- Control risks from manufacture, processing, distribution, use and disposal of electronic wastes.
- Encourage beneficial reuse of "e-waste" and encouraging business activities that use waste". Set up programmes so as to promote recycling among citizens and businesses:
- Educate e-waste generators on reuse/recycling options.

(iii) Governments must encourage research into the development and standard of hazardous waste management, environmental monitoring and the regulation of hazardous waste disposal.

(iv) Governments should enforce strict regulations against dumping e-waste in the country by outsiders: Where the laws are flouted, *stringent penalties* must be imposed. In particular, custodial sentences should be preferred to paltry fines, which these outsiders/ foreign *nationals can* pay.

(v) Governments should enforce strict regulations and heavy fines levied on industries, which do not practise waste prevention and recovery in the production facilities.

(vi) Polluter pays principle and extended producer responsibility should be adopted.

(vii) Governments should encourage and support NGOs and other organizations to involve actively in solving the nation's e-waste problems.

(viii) Uncontrolled dumping is an unsatisfactory method for disposal of hazardous waste and should be phased out.

(viii) Governments should explore opportunities to partner with manufacturers and retailers to provide recycling services.

(II) Responsibility and Role of Industries

1. Generators of wastes should take responsibility to determine the output characteristics of wastes and if hazardous, should provide management options.

2. All personnel involved in handling e-waste in industries including those at the policy, management, control and operational levels, should be properly qualified and trained. Companies can adopt their own policies while handling e-wastes. Some are given below:

- Use label materials to assist in recycling (particularly plastics).
- Standardize components for easy disassembly.
- Re-evaluate 'cheap products' use, make product cycle 'cheap' so that it has no inherent value that would encourage a recycling infrastructure.
- Create computer components and peripherals of biodegradable materials.

- Utilize technology sharing particularly for manufacturing and de-manufacturing.
- Encourage/promote/require green procurement for corporate buyers.
- Look at green packaging options.

3. Companies can and should adopt waste minimization techniques, which will make a significant reduction in the quantity of e-waste generated and thereby lessening the impact on the environment. It is a "reverse production" system that designs infrastructure to recover and reuse every material contained within e-wastes metals such as lead, copper, aluminium and gold, and various plastics, glass and wire. Such a "closed loop" manufacturing and recovery system offers a win-win situation for everyone. Less of the earth will be mined for raw materials, and groundwater will be protected, researchers explain.

4. Manufacturers, distributors, and retailers should undertake the responsibility of recycling/disposal of their own products.

5. Manufacturers of computer monitors, television sets and other electronic devices containing hazardous materials must be responsible for educating consumers and the general public regarding the potential threat to public health and the environment posed by their products. At minimum, all computer monitors, television sets and other electronic devices containing hazardous materials must be clearly labelled to identify environmental hazards and proper materials management.

(III) Responsibilities of the Citizen

Waste prevention is perhaps more preferred to any other waste management option including recycling. Donating electronics for reuse extends the lives of valuable products and keeps them out of the waste management system for a longer time. But care should be taken while donating such items, i.e. the items should be in working condition.

Reuse, in addition to being an environmentally preferable alternative, also benefits society. By donating used electronics, schools, non-profit organizations, and lower-income families can afford to use equipment that they otherwise could not *afford.*

E-wastes should never be disposed with garbage and other household wastes. This should be segregated at the site and sold or donated to various organizations.

While buying electronic products opt for those that:

- o are made with fewer toxic constituents
- o use recycled content
- o are energy efficient
- o are designed for easy upgrading or disassembly
- o utilize minimal packaging
- o offer leasing or take back options
- o have been certified by regulatory authorities. Customers should opt for upgrading their computers or other electronic items to the latest versions rather than buying new equipment.

8. Conclusions

E-waste is a matter of great concern for the entire world. Now the time has come to start composite efforts in the proper disposal method of e-waste. An international agency shall be created to work in the direction of R & D activity to minimize the hazardous effects of e-waste and developing the biodegradable electronic parts. The newly developed electronic parts shall be reclyable and reusable with no harmful effect on nature then only can we get rid of the problem of e-wastre from the world.

9. Bibliography

Books

Freeman, M.H. 1989. *Standard Handbook of Hazardous Waste Treatment and Disposal,* McGraw Hill Company, USA.

Third World Network. 1991. *Toxic Terror: Dumping of Hazardous Wastes in the Third World,* Third World Network, Malaysia.

Journals

Devi, B.S, Shobha S.V. and Kamble, R.K. (2004). E-Waste: The Hidden Harm of Technological Revolution, *Journal IAEM,* Vol. 31, pp. 196-205.

Ramachandra T.V. and Saira V.K. (2004). Environmentally Sound Options for Waste Management, *Envis Journal of Human Settlements*, March 2004.

Website

- www.ewastejournal.com/tag/ewaste management
- www.iimm.orglknowledgebank
- www.swlf.ait.ac.th

3

Green Marketing to Sustainable Environmental Development

Praveen Sahu, S.K. Singh,
Gaurav Jaiswal and Sweta Gupta

Introduction

Green marketing refers to marketing processes whereby the products or services are offered to the customer on the basis of their environmental benefits. Environmental benefits of the products or services that cover a wide range of activities could be any changes in manufacturing process, packaging process or any modification in the product itself. The concept of green marketing was propounded in the late 1980s and early 1990s. In 1975, the American Marketing Association (AMA) held its first workshop on ecological marketing. The workshop resulted in one of the first books on green marketing entitled *Ecological Marketing*. Green marketing is defined as "Green or environmental marketing consists of all the activities designed to generate and facilitates any exchanges intended to satisfy human needs and wants occurs, with minimum detrimental impact on the natural environment."

Green marketing has been successful in grabbing the attention of the customers who are concerned with environmental changes. It has been a cause of concern that "green marketing is a way to environmentally sustainable development". World Commission on Environment and Development defined sustainable development as meeting "needs of present without compromising the ability of future generations to meet their own need". Definition of green marketing incorporates that green products and services have minimum detrimental impact on the natural environment. Probably the minimum detrimental impact on environment today would be able to serve the needs of future generations and could lead to sustainable environmental development.

Opportunities to Marketers

In the past decade marketers have been successful in harnessing the public attention towards green marketing. Markets are full of products with environmental benefits. Eco-friendly products grab the attention of customers. Customers being concerned about the environmental changes search for such products. Marketers have been trying to lure the customers through green marketing in a number of ways. Companies are finding new innovative ideas for attaching green marketing to their products. Some such examples are

- McDonald's restaurant–their napkins, and bags are made of recycled papers.
- Badarpur Thermal Power Station of NTPC in Delhi is innovating new ways to utilize coal-ash, which has been a major source of pollution.
- Xerox Company has introduced a high quality recycled photocopier paper.

The market has imposed competitive pressure on firms that are not deploying green marketing. As the demand (for product/services) changes, it exerts pressure on firms to accept the change as an opportunity. Probably, the firms marketing the goods with environmental benefits will have a competitive advantage over the firms marketing the non environmental goods. When the number of firms accepting green marketing trends, making their products environment-friendly would reduce the burden on environment, i.e. definitely reduce the detrimental impact on environment and could pave the way towards sustainable environmental development.

Social Responsibility Towards Sustainable Environmental Development

Firms are now recognizing that they are a part of a community and they must take the responsibility towards the environment as a part of their corporate social responsibility. These must heal the environmental damages which they are causing. In the present global scenario where the environmental issues are paramount, the number of firms are adopting and incorporating environmental responsibility in their corporate culture. These are incorporating it by making their product environment-friendly or by taking the initiative and responsibility to protect the environment. For this, firms are investing a large chunk of money in activities like recycling, modifying packaging, etc. to minimize its deteriorating effect on environment.

This does not necessarily imply that all the firms who have undertaken environmental responsibility as a part of their corporate social responsibility have actually improved their performance. Some firms are misleading customers merely for increasing market share. While some are propagating their product as environment-friendly without considering the accuracy, efficiency and effectiveness of their product being environment-friendly.

There is a lack of standards about the greenness of the product which creates confusion among the customers. Despite all these, green marketing is continuously flourishing because of the growing global concern about the environmental changes.

Governmental Pressure Towards Environmental Sustainability

Environmental concern, has led all the governments: to think towards the environmentally sustainable development. All the governments want to protect society. This attitude of the government has important implications on green marketing. Government is trying to regulate the detrimental impact on the environment by utilizing various measures and environmental policies like –

1. Decentralized policies (enforcing liability laws, changes in property rights and voluntary actions).
2. Command and control policies (setting standards)
3. Incentive based policies (taxes and subsidies, transferable discharge permit).

All these steps taken by the government are forcing the marketers to incorporate environmental benefits associated with them. The government's attitude towards sustainable environmental development for mankind has enforced environmental marketing activities. Obviously, if measures are taken on a large scale it could lead to sustainable environmental development. This is because of government rules, regulations and policies towards environmental protection have caused most of the firms to divert their marketing towards the trend of green marketing. One such example is introduction of CNG in New Delhi. New Delhi was being polluted at an extremely high pace. Then the Supreme Court of India forced it to shift to alternative fuels. In the year 2002, a directive was issued by the government to completely adopt CNG in all the public transportation in New Delhi state to limit pollution.

Public Concern for Environmental Issues

In the present scenario, customers are very concerned about the environmental issues and are ready to pay the price for saving environment. People today are aware of the fact that we ourselves are the very cause of environmental degradation and are prepared to bear the responsibility, to heal and rejuvenate the environment. Steps are taken worldwide by NGOs and many such other societies to protect the environment. People are protesting all over the world against anti-environmental activities. In 1992 a study of 16 countries demonstrate that more than 50% of the consumers in each country expressed concern about the environment, except Singapore. More surveys show that people all over the world are modifying their behaviour because of the environmental concern. The attitude and behaviour of the customers have forced the firms towards green marketing. Customers are now seeking products which least affect the environment. It could be presumed that people are committed to their environment and understand their responsibility towards environment and are modifying their consumption behaviour accordingly which would lead to sustainable environmental development and future generations will not compromise.

Significance of Green Marketing

Sustainable environmental development is not an option but the need of the time. Today the world is facing environmental degradation from poverty, population and pollution. The poor depend on environment for their livelihood, population exerts pressure on natural resources for goods and services while pollution results from increased economic activities due to economic growth. Our environment renders security for present and future generations and any bad impact on the environment closely affects the well-being of the human race. Green marketing, i.e. marketing environment-friendly products and services, promoting environment specific technology and diverting existing economic activities towards environment specific technologies could only pave the way for sustainable environmental development.

Green marketing conceptualizes developing and promoting products and services so as to satisfy needs and wants of a customer at an affordable pricing, with least deterioration of the environment. Development of new and improved products and services fasten with environmental benefits provides firms to access the new market and increases their profitability and such firms enjoy competitive advantage over those which are not producing environment-friendly products. Companies are adopting green marketing for several reasons. Some of the reasons are — corporate social responsibility, competitive advantage, government pressure, customer pressure, competitive pressure and for profitability.

Natural resources are limited and human wants and needs are unlimited. Our resources are extinguishing at a very fast pace. And if this pace continues, our future generation will be deprived of. Therefore it is of utmost importance for marketers to utilize resources efficiently and effectively in order to provide sustainable and socially responsible products and services. This has made green marketing inevitable.

Problems Associated with Green Marketing

Green marketing is in its incipient stage and a lot of work has to be done to fully utilize the potential of green marketing. Most people are dazed about green marketing. Firms using green marketing must confirm that their activities, advertisements, products, etc. are not misleading to customers. Customers tend to be sceptical of green claims, as there are a number of green products and services available in the market without proper certification and labelling. Customers are compelled to rely on the manufacturer's marketing claims in the absence of proper certification and labelling standards.

Production of green products is costlier because green products require recyclable material. Cost associated with the green product is one of the important hurdles for green marketing. Using of recyclable material may require establishing recycling plant or buying recycled material from the market which may cost more than the fresh material available in the market.

Research and development for innovating new technologies for making green products requires huge investments. Many companies in underdeveloped and developing countries are unable to spend large chunks of money on R and D. So in these countries green marketing has been neglected for the economic aspect of marketing.

It is very difficult to evaluate the environmental gains and losses in monetary terms and without any pricing it is more difficult to incorporate environment into the economic system. All the economic instruments such as taxes, charges and subsidies which aims to curb environmentally damaging behaviour through environmental costs are very, difficult to evaluate.

Many of the customers are unwilling to pay high prices for green products. Income level of the customers and their financial securities are the decisive factors for willingness to pay. It is challenging for companies to lure customers for high priced green products.

Conclusion

It is a well known fact that sustainable environmental development is the need of the time. Governments are trying to achieve sustainable environmental development through rules, regulations and policies for the sake of mankind. Marketers are also recognizing their responsibility towards the environment and headed towards green marketing. It is not the sole duty of the governments or marketers to preserve the environment. Undeniably it is the people who are responsible for the deterioration of the environment. It is the people who demand goods and services and the marketers serve to satisfy their needs and wants. Sustainable environmental development could be achieved by the combined efforts of the governments, marketers and customers. Unless individuals are committed to their environment and do not change their consumption behaviour accordingly, it is very difficult to attain sustainability in the environment. Government should intervene and ensure that individuals must pay for cleaner environment and for high priced goods and services offered by green marketers because people themselves are the ultimate cause of environmental degradation. Companies should also be committed to the environment. These should not only provide environment-friendly goods and services but should also pressurize their suppliers to modify their activities and to behave environmentally. The economic activities must be diverted towards environment specific and resource saving technologies. Environment should be preserved by law enforcement, technological improvements, providing economic incentives and by public participation.

REFERENCES

A Conceptual Framework on Green Marketing by Assistant Professor Rohit Kumar Mishra, Institute of Management & Information Science.

Green Marketing: Opportunities and Challenges by Prof. Sanjit Kumar Dash *http;//www.coolavenues.com/know/mktg/sanjit-green-marketing-4.php*

www.greenmarketing.net/stratergic.html

www.epa.qld.gov.au/sustainabl_industries

www.wmin.ac.uk/marketing research/marketing /greenmix.html

http://www.indiaenviromentportal.org.in/node/4244

A Look Back at Green Marketing in 2007
http://marketinggrreen.wordpress.com/2007/12/29/a look back at green marketing in-2007/

The Tradable Permits Approach to Protecting the Commons: What Have We Learned? By Tom Tietenberg, Mitchell Family Professor of Economics.
http://www.colby.edu/personal/thtieten/

Sharon Beder, "The Environment Goes to Market', *Democracy and Nature* 3(3), 1997. pp. 90-106.

4

Global Warming: The Last Phase to Save the Earth

K.S. Rathod, Prashant Sahu and Ramesh Kumar

Introduction

"Here the world to avoid the rising menace of global warming is the last opportunity" Representative of 192 countries, Copenhagen *Summit, December 7, 2009.*

Global warming is real but not due to greenhouse gases. The sun is going through its cycles as it always has. The earth has been colder than it is now and much warmer based on the ice core samples. In fact, overall CO_2 levels lag the temperature at a rate of about 400 years. The temperature rising and falling actually drive CO_2 levels. The total CO_2 levels contributed by man are small compared to what the volcanoes, animals, vegetation and the oceans produce. In fact using CO_2 levels as an indicator that the planet is heating up is like saying that the leaves falling off the tree are bringing about winter.

Environment is a very broad concept. Everything that affects us during our lifetime is collectively known as environment. As human beings we are often concerned with surrounding conditions that affect people and other organisms. Today, all over the world there is growing concern about the deteriorating quality of environment and efforts are being made to stop the widespread abuse of environment and improve its quality.

Environment is a broad concept and further it may be categorized in different forms like political, social, animal, economical, demographic, legal, geographic, etc. which is affecting the human life directly or indirectly across the world. In this connection global warming already disrupts millions of lives daily in the forms of destructive weather patterns and loss of habitat. What is already happening is only the tip of the melting iceberg, for it is our children and grandchildren who may suffer most from the effects of global warming. Hundreds of millions of people may be exposed to famine, water shortages, extreme weather conditions and a 20- 30% loss of animal and plant species if we do not reduce the rate of global warming and reduce GHG emissions. On the other hand, having warmer winters means longer growing seasons in temperate and subarctic climes, sometimes allowing an additional crop to be planted and harvested each year, or simply making the existing crops more productive.

Reducing carbon and greenhouse gas emissions will not only make our personal living space more sustainable but it will also save our money in both the short and long term. Global warming is occurring more rapidly than it was originally expected to only forty years ago, the major worry was global cooling. Even if we remain cynical, however, and disagree with the consensus of scientists, will result in benefit from reduced pollution, a more healthful lifestyle and increased savings from enacting these simple activities that will not reduce the quality of your life.

The earth has warmed up by about 0.6°C in the last 100 years. During this period, man made emissions of greenhouse gases have increased, largely as a result of the burning of fossil fuels and deforestation. In the last 20 years, concern has grown that these two phenomena are, at least in part, associated with each other. That is why I took into consideration the aspect of global warming in this paper and will try to highlight all the possible outcomes to save the interests of all human beings across the world. This is now considered most probably to be due to the enhanced greenhouse effect. The paper outlines some ways that we can act to help prevent the earth from warming further with help of suggesting and analysing the causes and effects. While humankind has the ability to destroy the planet, we can also help protect and sustain it.

Definition of global warming. What is global warming?

Global warming is the observed and projected increase in the average temperature of earth's atmosphere and oceans.

Objective of the Paper

1. Explain the concept of environment and global warming
2. Describe the various facets of human environment interaction to affect global warming
3. To find out cause and effect of global warming and
4. To find out the impact of global warming on environment
5. To provide possible outcomes to reduce global warming and to save environment
6. To provide few suggestions to control global warming

Research Methodology

The following methodology used in the terms of research work

1. To select the universe on which the research work will be done.
2. Research design will be framed on the basis of the problems regarding the global warming
3. The study will be based on secondary data. The secondary data will be collected from published material from different industries reports, IPCC, newspaper articles and journals published by eminent professors, readers and international conferences.
4. Collected data will be classified in tabulation form.
5. Research plan: Research type: Quantitative exploratory research.

Cause of Global Warming

Almost 100% of the observed temperature increase over the last 50 years has been due to the increase in the atmosphere of greenhouse gas concentrations like water vapour, carbon dioxide (CO_2), methane and ozone. Greenhouse gases are those gases that contribute to the greenhouse effect (see below). The largest contributing source of greenhouse gas is the burning of fossil fuels leading to the emission of carbon dioxide

The *effects of global warming* and climate change are of concern both for the environment and human life. Evidence of observed climate change includes the instrumental temperature record, rising sea levels, and decreased snow cover in the Northern Hemisphere. J31 According to the IPCC Fourth Assessment Report, "[most] of the observed increase in global average temperatures since the mid 20th century is very likely due to the observed increase in [human greenhouse gas] concentrations". It is predicted that future climate changes will include further global warming (i.e. an upward trend in global mean temperature), sea level rise, and a probable increase in the frequency of some extreme weather events. Ecosystems are seen as being particularly vulnerable to climate change. Human systems are seen as being variable in their capacity to adapt to future climate change. To reduce the risk of large changes in future climate, many countries have implemented policies designed to reduce their emissions of greenhouse gases.

Glaciers

Glaciers are among the most sensitive indicators of climate change advancing when climate cools (for example, during the period known as the Little Ice Age) and retreating when climate warms. Glaciers grow and shrink, both contributing to natural variability and amplifying externally forced changes.

Vegetation

A change in the type, distribution and coverage of vegetation may occur given a change in the climate; this much is obvious. In any given scenario, a mild change in climate may result in increased precipitation and warmth, resulting in improved plant growth and the subsequent sequestration of airborne CO_2. Larger, faster or more radical changes, however, may well result in vegetation stress, rapid plant loss and desertification in certain circumstances.

Ice Cores

Analysis of ice in a core drilled from an ice sheet such as the Antarctic ice sheet, can be used to show a link between temperature and global sea level variations. The air trapped in bubbles in the ice can also reveal the CO_2 variations of the atmosphere from the distant past, well before modern environmental influences. The study of these ice cores has been a significant indicator of the changes in CO_2 over many millennia, and continues to provide valuable information about the differences between ancient and modern atmospheric conditions.

Dendroclimatology

Dendroclimatology is the analysis of tree ring growth patterns to determine the age of a tree. From a climate change viewpoint, however, Dendroclimatology can also indicate the climatic conditions for a given number of years. Wide and thick rings indicate a fertile, well watered growing period, while thin, narrow rings indicate a time of lower rainfall and less than ideal growing conditions.

Pollen Analysis

Palynology is the study of contemporary and fossil palynomorphs, including pollen. Palynology is used to infer the geographical distribution of plant species, which vary under different climate conditions. Changes in the type of pollen found in different sedimentation levels in lakes, bogs or river deltas indicate changes in plant communities which are dependent on climate conditions.

Insects

Remains of beetles are common in freshwater and land sediments. Different species of beetles tend to be found under different climatic conditions. Given the extensive lineage of beetles whose genetic makeup has not altered significantly over the millennia, knowledge of the present climatic range of the different species, and the age of the sediments in which remains are found, past climatic conditions may be inferred.

The main cause of global warming is our treatment of nature

- Why have warnings about climate change been ignored for more than 20 years?
- Why were ever more scientific evidence demanded to find the coherence of man made C02 emissions as cause of global warming?
- Why wasn't common sense reason enough to act?
- Why can one still today find people who stick their head in the sand and don't want to understand what's going on in the earth's atmosphere?
- Why do most people refuse to change their personal behaviour voluntary in order to reduce CO_2 emissions caused by their activities?

Effects of Global Warming

The Greenhouse Effect

When sunlight reaches earth's surface some heat is absorbed and warms the earth and most of the rest is radiated back to the atmosphere at a longer wavelength than the sun light. Some of these longer wavelengths are absorbed by greenhouse gases in the atmosphere before they are lost to space. The reflecting back of heat energy by the atmosphere is called the "greenhouse effect". The major natural greenhouse gases are water vapour, which causes about 36-70% of the greenhouse effect on earth (not including clouds); carbon dioxide CO_2, which causes 9-26%; methane, which causes 4-9%, and ozone, which causes 3-7%. Other greenhouse gases include, but are not limited to, nitrous oxide, sulfur hexafluoride, hydro fluorocarbons, perfluorocarbons (A powerful greenhouse gas emitted during the production of aluminium) and chlorofluorocarbons. The emission of carbon dioxide into the environment mainly from burning of fossil fuels (oil, gas, petrol, kerosene, etc.) has been increased dramatically over the past 50 years, see graph below.

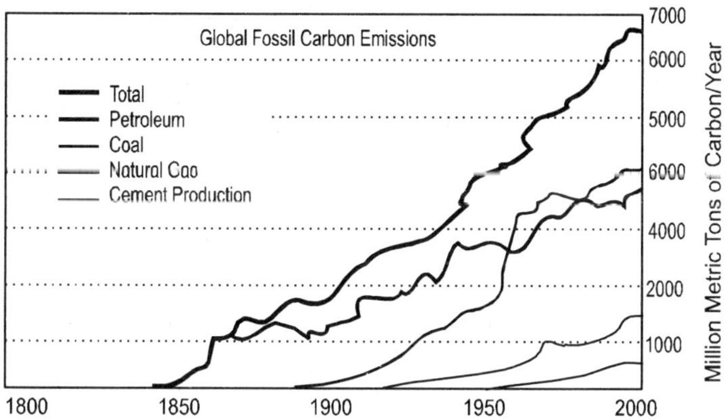

Fig. 1: Cause for global warming: Carbon dioxide emissions in million tons per year over the last 200 years (Graph taken from http://www.globalwarmingart.com/wiki/ Image:Global_Carbon_Emission_by_Type_png)

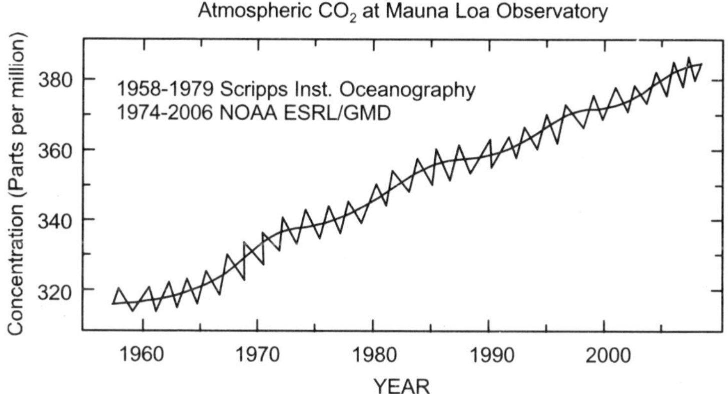

Fig. 2.: Global warming cause: Concentration of carbon dioxide has dramatically increased in the last 50 years

(Source: NOAA, National Oceanic and Atmospheric Administration*)*

Global Trends in Major Greenhouse Gases to 1/2003

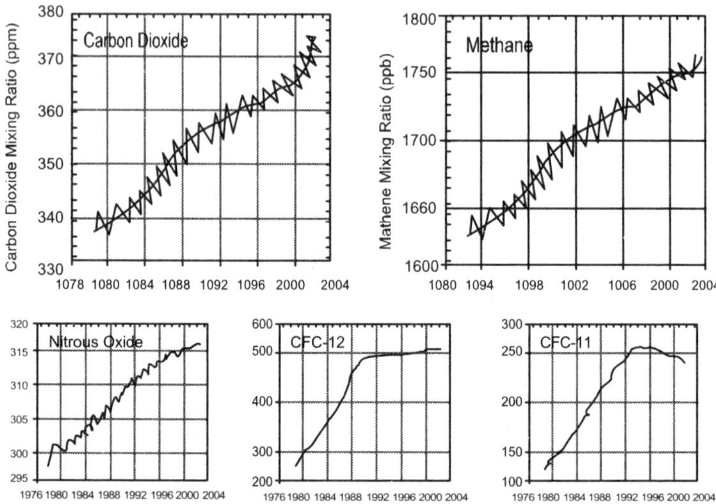

Global trends in major long-lived greenhouse gases through the year 2002. These five gases account for about 97% of the direct climate forcing by long-lived green-house gas increases since 1750. The remaining 3% is contributed by an assortment of 10 minor halogen gases, mainly gases, mainly HCFC-22, CFC-113 and CCl_4.

Fig. 3: Trends for greenhouse gases: Carbon dioxide (CO_2) and nitrous oxide (NO_x) ***concentrations in the atmosphere are still increasing.*** *For the other major greenhouse gases, the steady upward trend has been broken. (Graph taken from http://upload.wikimedia.org/wikipedia/en/b/bb/Major_greenhouse _gas_trends.png)*

Annual Grreenhouse Gas Emissions by Sector

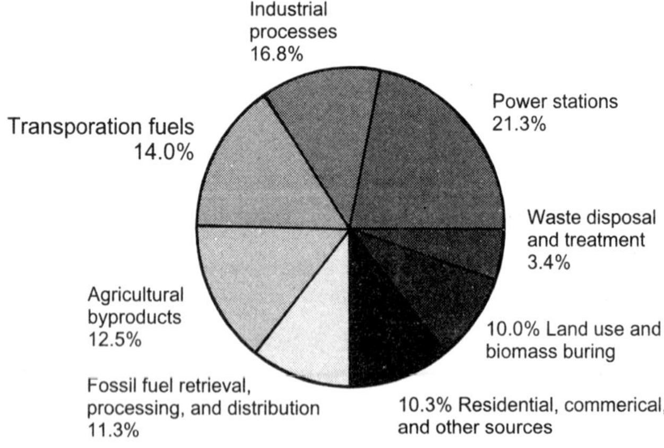

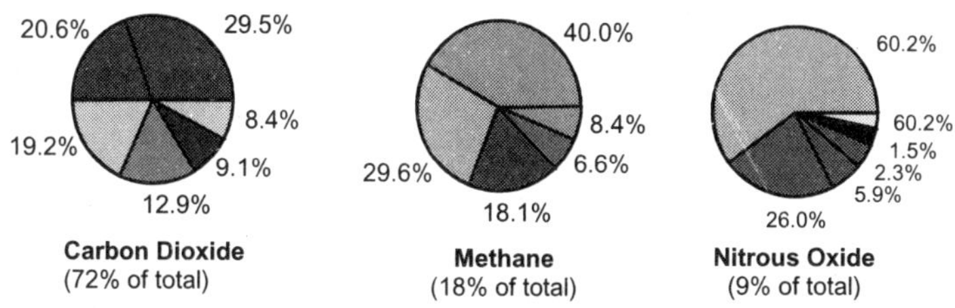

Fig 4: From which sectors do the major greenhouse gas emissions come from? The lower part of the picture shows the sources individually for the gases carbondioxide, methane and nitrous oxide, respectively. (Graph fromhttp://en.wikipedia.org/wiki/Image:Greenhouse_Gas_by_Sector.png)

The increase of greenhouse gas concentration (mainly carbon dioxide) led to a substantial warming of the earth and the sea, called global warming. In other words: The increase in the man-made emission of greenhouse gases is the cause for global warming. For the effects of global warming see above.

In listed above There are two major effects of global warming?

- Increase of temperature on the earth by about 3° to 5° C (5.4° to 9° Fahrenheit) by the year 2100.
- Rise of sea levels by at least 25 metres (82 feet) by the year 2100.

It took more than 20 years to broadly accept that mankind is causing global warming with the emission of greenhouse gases. The drastic increase in the emission of C02 (carbon dioxide) within the last 30 years caused by burning fossil fuels has been identified as the major reason for the change of temperature in the atmosphere.

More than 80% of the world wide energy demand is currently supplied by the fossil fuels coal, oil or gas. It will be impossible to find alternative sources, which could replace fossil fuels in the short or medium term. The energy demand is simply too high.

Effects of Climate Change on Asia

- By the 2050s, freshwater availability in Central, South, East and South East Asia, particularly in large river basins, is projected to decrease
- Coastal areas, especially heavily populated mega delta regions

in South, East and South-East Asia, will be at greatest risk due to increased flooding from the sea and, in some mega deltas, flooding from the rivers

- Climate change is projected to compound the pressures on natural resources and the environment, associated with rapid urbanization, industrialization and economic development
- Endemic morbidity and mortality due to diarrhoeal (regular) disease primarily associated with floods and droughts are expected to rise in East, South and South East Asia due to projected changes in the hydrological cycle

Prediction for future temperature increase (global warming predictions)

According to different assumptions about the future behaviour of mankind, a projection of current trends as represented by a number of different scenarios gives temperature increases of about 3° to 5° C (5° to 9° Fahrenheit) by the year 2100 or soon afterwards. A 3°C or 5° Fahrenheit rise would likely raise sea levels by about 25 metres (about 82 feet).

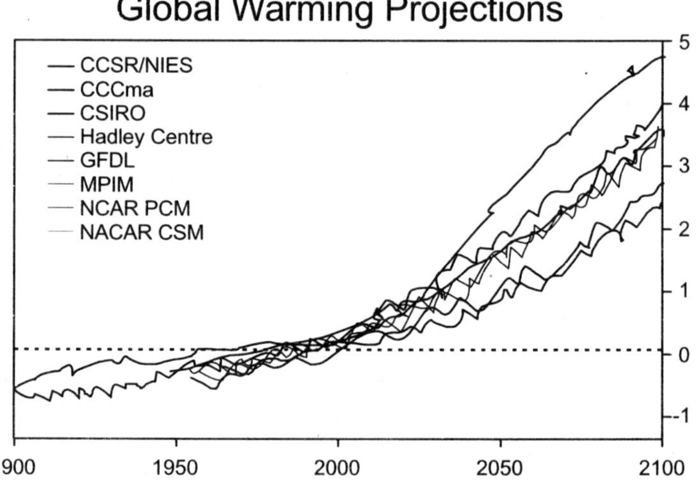

Fig 4: Definition for global warming: Temp. increase until the year 2100 (graph from http://www.globalwarmingart.com/wiki/Image: Global_Warming_ Predictions_png.

Suggestions

1. **Action by the government** – It is a must to be understood by the governments all over the world that they need to take strict action to reduce the CO_2 (carbon dioxide) which is mostly affecting global warming. It is needed to implement the policy to cover the gas emission and also the need to control and effective monitoring of the policy. In this connection, recently held at the Copenhagen climate negotiations have arranges on 7th December and discussed by various Prime Ministers of the world including our Prime Minister Manmohan Singh Monday to meet out cretin agreement to control global warming and to save the future of our grandsons.
2. **Implementation of advertisements** – Governments of the world need to made emotional appeal to the general public to stop production of CO_2 from different sources and must make advertisement hoardings, awareness programmes, tele film, nukkad natkak, pamphlets, etc. as done in case of polio disease by different governments in the world.
3. **Social responsibility of the companies** – As we all know that all over the world lacks industries are operating and also producing dangerous CO_2 and others gases which harm not only human beings but most essentially the environment which in turn is in the form of global warming. That is why it is must to prevent those companies who producing the carbon more then its capacity due to not mentioned machinery. It is the duty of the companies to consider these alarming situations as CSR (Corporate Social Responsibility) and need to revive production, machinery, technology and most important is attitude, in favour of nature and the world in which they are earning profit.
4. **Duties of film industries** As we know that films allays the impact on the human mind and attitude that is why it is the duty of the film maker to make involve those story which are associated or based on climate or nature. By which people of the world's make understand the reality of environment

and think over the issue which can save our world. A thousand films are being produced across the world based on disasters, cyclones, earthquakes, etc. The recently produced film 2012 impacting on the human mind as the world is going to finish in 2012 but this film also missed the message by which this 2012 was made, that is "*Why is the world ending in 2012?. They need to specify the reasons due to which this disaster is taking place. (Based on Maya civilization calendar)*

5. **Inclusion of one subject** of global warming in academic curriculum = Now it is the time to think of how we can develop a positive attitude of the people to save the earth because people in the world start their thinking process from their childhood and this is the best time to feed the memory since beginning to add curriculum in the book which will ultimately turn in the positive attitude to develop children's thinking process.
6. **As we know very well that** the carbon dioxide emissions (carbon footprint) caused by our personal behaviour is driven to a large extent by the type and quality of our food. The amount of greenhouse gases caused by the production of food differs very much from one food type to the other (see table below). Worst is meat and in particular beef. An environmental-friendly and "climate change-friendly" nourishment must be adopted as follows:
    - Very little (or no) meat
    - Eat primarily organic food
    - Seasonal food is preferred
    - Regionally produced food is preferred

Food Group	*Food*	*CO_2-Emissions (in gm per kg food)*
Meat and sausages	Beef	13,300
	Raw sausages	8,000
	Ham (pork)	4,800
	Poultry	3,500
	Pork	3,250

	Butter	23,800
	Hard cheese	8,500
	Cream	7,600
Milk and	Eggs	1,950
dairy products	Quark curd	1,950
	Farmer cheese	1,950
	Margarine	1,350
	Yogurt	1,250
	Milk	950
Fruits	Apples	550
	Strawberries	300
Baked goods	Brown bread	750
	White bread	650

Example: The production of I kg beef causes about 13.3 kg of CO_2. The same quantity of CO_2 is released when we burn about 6 litres of petrol!

7. **Implementation of legislation**– Environmental regulation is seen to play different roles during different phases of a country's development. Initially, environmental regulation serves a reactive purpose as a means of cleaning up after the new technologies and the new industries. During this initial phase of development it is often seen that growth is accompanied by increasing environmental degradation, and it is often assumed to be the case that environmental degradation is the obvious and unavoidable outcome from a process of economic growth. That is why governments of different countries need to take strict action to watch over and control the CO_2 and other harmful gases to save the earth.
8. **Responsibility of Media**– In today's era media is a powerful and strong source of information to communicate messages from one place to another in less than a minute and in this connection it becomes the moral duty to them to made some necessary message and spread it, to general public on daily basis so as to create awareness among masses to safe the earth. The message must be a bundle of different information who causes for global warming and soon the

time will come when we will receive the news that CO_2 level have reduced in compared to last year but this process must not be 'stopped till to achieve our goal.

9. **Use Public Transport–** Public transport, public transportation, public transit or mass transit comprises all transport systems in which the passengers do not travel in their own vehicles.

   While it is generally taken to include rail and bus services, wider definitions would include scheduled airline services, ferries; taxicab services etc. any system that transports members of the general public. A further restriction that is sometimes applied is that it should take place in shared vehicles, which would exclude taxis that are not shared-ride taxis.

10. **Use Renewable Energy Like Wind Power–** One means of reducing carbon emissions is the development of new technologies such as renewable energy such as wind power. Most forms of renewable energy generate no appreciable amounts of greenhouse gases except for bio fuels derived from biomass.

11. **Use Smart Cooler, Heater and Air Conditioner–** About half the energy we use in our homes goes to heating and cooling. Changing air filters annually, having our system checked annually and using a programmable thermostat are all easy things we can do. Just by using a programmable thermostat, you can save about 1,800 pounds of carbon dioxide a year and about $100 (Rs. 5000) a year in energy costs.

12. **Tune up and maintain vehicles properly–** Unscientific maintain of vehicle leads to environment pollution. Vehicles; regardless of category are increasing day by day' all over the world. The smoke released by these vehicles damage ozone layer. But it is impossible to stop the arrival of new vehicles. What can be done to the maximum is, to maintain the vehicles properly. Adopting scientific method to maintain our loved cars and bikes will play predominant role in controlling global warming.

13. **Clean the air in your house–** Cleaning the air inside the house is the most important thing. By doing so you will automatically contribute for global warm control. There are many things you can do to clean our house. Use proper vacuum cleaner for the purpose. Clean regularly and continuously. Put dust avoiding curtains and use houseplants. Do not keep the dustbin unchecked. Even take maximum care while dispatching waste materials. Try to grow as much as saplings (young tree) inside our compound.
14. **Reduce electricity usage to the maximum–** Switch off unwanted electric equipments immediately. Or do not use them if not necessary. Often we find shining tube, running fan, running TV ... etc. One may be sound enough to pay the electric bill at the end of the month, but what about the energy that has been wasted? Replace the old ones with energy efficient lighting. Also, improve the efficiency of home appliances. If not possible, go for an energy saving appliances.
15. **Prefer recycling–** Preferring reusable products instead of disposables will help in reducing the waste. When you buy a product, make sure that the packing is quite reasonable one. In other words, packing should not exceed the size of the product. Always try to recycle household waste. By recycling the household waste, one can save 2,400 pounds of carbon dioxide annually. Here both the ,entrepreneur's' and public should join hands together for a cause. Always try to educate others on preferring recycling products.
16. **Forestation–** First we have to stop cutting the forest whatever is remaining now. When it stops completely, then we can take measures to increase the density of forests. It may includes growing trees in their campus and their agricultural lands of every individual with their own interest. Definitely it will some how reduce the emission of harmful gases like CO2.
17. **Responsibility of individual and family as a whole** never but not the least, it is not only responsibility of the

government but also for the individual belong to different age group, sex , education, caste and community as whole to control the greenhouse effects which is destroying our earth.

18. **Need to work together–** It is needed to make a unique policy by various countries of the world that to work together to stop C02 emission which are really a subject matter to cause Global Warming.

Conclusion

According to the International Panel on Climate Change (IPCC) if we are to have any chance of stopping Anthropogenic Global Warming/Anthropogenic Global Climate Change, we must reduce the worldwide carbon emissions to less than one tenth of what they are today. That means that we must reduce worldwide carbon emissions to less that nine tenths of what they are today. The only way to achieve a reduction that large is to ban the use of fossil fuels worldwide, make legislation, plantation, individual efforts as listed in the suggestion, stop eating meat, etc. That means that we must ban all forms of motorized transportation. We must ban the heating of homes and we must ban the heating of water that we use for bathing and cooking. We know that it is not possible with 100% but if we do not start from 1%, we can not secure the lives of the future generation.

If we are serious about stopping Global Warming/ Climate Change it appears that we will be taking a lot of cold showers, cyclones, earthquakes and non-diagnosable diseases.

Bibliography

Abarbanel, Albert, and Thomas McCluskey (1950). "Is the World Getting Warmer?" *Saturday Evening Post, 1 July, pp.* 22 23, 57 63.

Abbot, Charles G., and F.E. Fowle, Jr. (1908). "Income and Outgo of Heat from the Earth, and the Dependence of Its Temperature Thereon." *Annals of the Astrophysical Observatory (Smithsonian Institution, Washington DC)* 2: 159 176.

Abelmann, Andrea, et al. (2006). "Extensive Phytoplankton Blooms in the Atlantic Sector of the Glacial Southern Ocean." *Paleoceanography* 21: PA1013 [doi: 10.1029/2005PA001199, 2006].

Abelson, P.H. (1977). "Energy and Climate." *Science* 197: 941.

Adler, Jerry (2007). "Moment of Truth." *Newsweek* (April 16), pp. 45-48.

Agrawala, Shardul (1998). "Context and Early Origins of the Intergovernmental Panel for Climate Change." *Climatic Change* 39: 605-20.

Agrawala, Shardul (1998). "Structural and Process History of the Intergovernmental Panel for Climate Change." *Climatic Change* 39: 621-42.

Agrawala, Shardul (1999). "Early Science Policy Interactions in Global Climate Change: Lessons from the Advisory Group on Greenhouse Gases." *Global Environmental Change* 9(2): 157-69.

Andreae, Meinrat O., et al. (2005). "Strong Present Day Aerosol Cooling Implies a Hot Future." Zwally, H.Jay, et al. (2005). "Mass Changes of the Greenland and Antarctic Ice Sheets and Shelves and Contributions to Sea Level Rise: 1992 2002." *J. Glaciology* 51: 509-27.

Chanchal Singh, Akansha (2007)."Global Warming and Climatology", xiv, 318 p, ISBN: 81-8370-121-1

Website

http://www.ias.ac.in/currsci/jul102009/9.pdf

http://www.merinews.com/article/global-warming humans-are-to-blame/15773854.shtml.

http://www.globalwamiingart.com

http://en.wikipedia.org

http://www.globalwanningart.com

http://en.copl5.dk

www.bhaskar.com

www.economictimes.indiatimes.com

http://en.copl5.dk

5

Recycling E-waste: Hazards, Compulsion and Safety

Praveen Sahu, Neha Mathur and Sanjay Shrivastava

Introduction to the Concept

The ongoing technological advancement has lead to the obsolescence of many of the electronic products within a very short period of time that is developing into a large surplus of unwanted electronic products or "e-waste." The disposing of e-waste in landfills has the potential to cause severe human and environmental health impacts.

The problem has become an issue of major concern, and thus a measure has been taken that has resulted in the growth of the electronic waste recycling business which is spreading as a consolidating business in all areas of the developed world. In the recent years, the electronic waste processing system has matured, followed by increased regulatory, public, and commercial scrutiny and a commensurate increase in the entrepreneurial interest as well. Part of this evolution has involved greater diversion of electronic waste from energy intensive down cycling processes (e.g. conventional recycling), where equipment is reverted to a raw material form.

This diversion is achieved through reuse and refurbishing/renovation. In the realm of solid waste, reuse means the equipment is still working and can be sold or donated, thus continuing the "life" of the product. Reuse is beneficial to the environment and ultimately to the society in the sense that it includes diminished demand for new products and virgin raw materials; larger quantities of pure water and electricity that are required for manufacturing; less packaging per unit; availability of technology to wider sections of society owing to greater affordability of products; and diminished use of landfills. Some organizations are involved in both reuse and recycling while others focus on one activity.

In developing or under developed countries like India, major cities such as Delhi, Mumbai and Bangalore are the extensive source of e-waste. Here the complex e-waste handling infrastructure is based on *waste recycling* and is operated by a very entrepreneurial sector. Even in India the reuse of materials or the extraction of secondary raw material has given rise to new businesses and with the help of rag pickers the waste dealers have easily adopted the new stream of waste management. It reflects that the e-waste recycling system is purely market-driven.

India being a developing country needs simpler, low cost technology keeping in view maximum resource recovery in environmental-friendly methodologies. Electronic waste or e-waste is one of the rapidly growing environmental problems of the world. In India, the electronic waste management assumes greater significance not only due to the generation of our own waste but also dumping of e-waste particularly computer waste from the developed countries. It is due to the lack of governmental legislation on e-waste, standards for disposal, proper mechanism for handling these toxic hi-tech products, mostly end up in landfills or partly recycled in a unhygienic conditions and partly thrown into waste streams.

This research article is thus framed to showcase the problem of e-waste before the society, the issues that are related to management of e-waste, the measures that are been taken till date to control e-waste and what other steps and possible solutions can be there keeping in view the prior suggested and applicable solutions, which are already in practice.

Meaning and Concept of E-waste

E-waste is a popular, informal name for electronic products nearing the end of their "useful life." Computers, televisions, VCRs, stereos, copiers, and fax machines are common electronic products. Many of these products can be reused, refurbished, or recycled.

The term "e-waste" is loosely applied to consumer and business electronic equipment that is near or at the end of its useful life. There is no clear definition for e-waste; for instance, whether or not items like microwave ovens and other similar "appliances" should be grouped into the category has not been established.

Certain components of some electronic products contain materials that render them hazardous, depending on their condition and density. For instance, California law currently views nonfunctioning CRTs (cathode ray tubes) from televisions and monitor as hazardous. All secondary computers, entertainment device electronics, mobile phones, refrigerators, television, etc. whether sold donated or discarded by their original owners comes under electronic waste. The residue or material that is repairable and can be reused or recycled but is :dumped or disposed or discarded by The buyer rather than recycled includes 'waste'. The *United States Environmental Protection Agency* (EPA) includes discarded CRT monitors in "hazardous household waste". But considers CRTs set aside for testing to be commodities if they are not discarded, speculatively accumulated, or left unprotected from weather and other damage.

Note that: Electronic waste, e-waste, e-scrap, or Waste Electrical and Electronic Equipment (WEEE) describes loosely discarded, surplus, obsolete, broken, electrical or electronic, devices.

Affect of E-waste on Human Life and Environment

On one side if e-waste contains valuable materials like gold, palladium, silver and copper it also consists of harmful and toxic substances like lead, cadmium and mercury, which when recycled needs to be taken good care of, even needs proper techniques/ methods with subsequent measures to recycle or else the release of toxic substances from the waste to be recycled can cause serious damage to the environment and human life as well:

These toxic substances can cause serious health problems which may last for generations and thus comes with heredity.

The rate of obsolesce of personal computers is very high, i.e. one in every two years. It's a general tendency of a consumer to get a new computer instead of getting the old computers upgraded, as they find it more convenient and it goes as per the need for the latest technology and configuration. Apart from this the attractive offers from the manufacturers and basically the cheap and affordable prices offered by the retailers and manufacturers makes a consumer buy a new system. This is again leading towards generating a huge amount of e-waste as nearly 70% of the literate population has become techno-friendly.

Since there are no such legislation of government on recycling of e-waste and handling of such highly toxic substances, the dumping of the electronic garbage ends up in landfills or sometimes they are partly recycled in an unhygienic conditions and or partly thrown into waste streams. Land filling E-waste is one of the most widely used methods of disposal and this is prone to hazards due to leachate that contains heavy water resources.

Note: *Landfills are underground facility where all the waste produced on planet are dumped and thus sealing it up in an engineered way that it doesn't seep through the air or the ground. They are the poisonous Pandora's Box.*

There are several examples relating to this, for instance; Cathode Ray Tube (CRT) has high contents of carcinogens such as lead, barium, phosphor and other heavy metals. If these CRTs are broken, recycled or disposed of in an uncontrolled environment in the absence of any safety precautions/measures, can release toxins in the air, soil and ground water and probably harm the workers.

Another example is the dangerous process which is recycling of halogenated chlorides& bromides that are used as flame retardants in plastics and forms persistent dioxins and furans on combustion at low temperatures (600-800 degrees centigrade).

Apart from this in Printed Circuit Boards and cables the presence 'of 'copper acts as a catalyst for dioxin formation when flame retardants are incinerated. The concentration of dioxins in the surroundings of open burning places generated from burning of printed wire boards has reached to an alarming situation as it is reaching 30 times the Swiss Guidance level. These toxins generate an increased risk of cancer if inhaled by workers and local residents or by entering the food chain via crops from the surrounding fields.

Sources for E-waste

(1) Cathode Ray Tube containing devices (CRT devices);
(2) Cathode Ray Tubes;
(3) Cathode Ray Tube containing computer monitors;
(4) LCD-containing laptop computers;
(5) LCD-containing desktop monitors;
(6) CRT-containing televisions;
(7) LCD-containing televisions (**excluding LCD projection televisions**);
(8) Plasma televisions (**excluding plasma projection televisions**); and
(9) Portable DVD players with LCDs.

1. Substances found in large quantities include epoxy resins, fiberglass, PCBs; PVC, thermosetting, plastics, lead, tin, copper; silicon; beryllium, carbon, iron and aluminum.

2. Elements found in small amounts include cadmium, mercury; and thallium.

3. Elements found in trace amounts include americium, antimony; arsenic, barium, bismuth, boron, cobalt, europium, gallium, germanium, gold, indium, lithium, manganese, nickel, niobium, palladium, platinum; rhodium, ruthenium, selenium, silver, tantalum, terbium, thorium, titanium, vanadium, and yttrium.

Rapid change in the technology; low initial cost, and planned obsolescence has resulted in a fast growing surplus of electronic waste around the globe.

There are several technical solutions available, but absence of government regulations and legislations also the least concern for implementation of such technologies makes it a weak concept.

An estimated 50 million tones of e-waste are produced each year. USA alone discards 30 million computers every, year.. Europe: disposes 100 million phones each year. An estimated 70% of heavy metals in the United States found in landfills arise from discarded electronics. This percentage is confined to only 2% in America.

Other major sources of e-waste are; equipment like sodium lamps, fluorescent tubes, VCR/DVD/CD players, radios, drills, lawn mowers, electric saws, sewing machines, coin slot machines, surveillance equipment, electric train sets, etc. The list is truly exhaustive. In broader sense, these waste products are called WEEE (Waste Electrical and Electronic Equipment

Involvement of Government in Management E-waste and Universal Waste Handlers

Starting next fiscal, consumers in India will no more be required to bother about managing their electronic and electrical equipment at the end of the lifecycle. The Centre is in the process of bringing out a legislation based on the draft proposals submitted by industry bodies and green campaigners including Greenpeace, which, for the first time will make the producers (manufacturers) responsible for the management of electronic waste (e-waste).

The new legislation which the government has agreed to approve by March 2010, will now make the 25 odd PC manufacturers and sellers who control about 75 per cent of the market share in the organized space, to implement take back policy for their end-consumers and recycle the same in an environment friendly manner. Besides, the unorganized players who hold about 25 per cent share in the PC market in India will also have to comply with the rule.

A brainchild of global environmental NGO Greenpeace, the draft proposals on e-waste management were formally approved by various stakeholders including the electronics goods manufacturers, industry body Manufacturers' Association for Information Technology (MAIT) and not for profit', organizations 'including Toxics 'Link,' GTZ and Greenpeace in an event held in Bangalore in April 2008.

"The government is currently in the process of developing a

dedicated set of rules which would govern the management and handling of electronic waste. These will be put in the public domain for comments by March 2010," Saroj, Director, Ministry of Environment and Forests.

E-Parisaraa, an eco friendly recycling unit on the outskirts of Bangalore which is located in Dobaspet industrial area, about 45 kms north of Bangalore, makes full use of e waste. E Parisaraa has developed a circuit to extend the life of Tube Lights. The circuit helps to extend the life of fluorescent tubes by more than 2000 years.

E-Parisaraa's Director Mr. P. Parthasarathy, an IIT Madras graduate, and former consultant for a similar e-waste unit in Singapore, has developed an eco friendly methodology for **reusing, recycling and recovery of metals**; glass and plastics with non-incineration methods. The hazardous' materials are segregated and sent for secure landfill for example phosphor coating, LED's mercury, etc.

The Ministry of Environment & Forests (MoEF), Government of India, as the highest national authority with regard to environmental legislation provides the policy and legal inputs. The Central Pollution Control Board (CPCB), an autonomous body under the (MoEF) is involved in developing the draft legislation as well as providing technical inputs at the national as well as state level.

DTSC has also adopted regulations (Chapter 23 of Title 22 of the California Code of Regulations) designating e-wastes as universal wastes. Because they pose lower immediate risk to people and the environment when properly managed, universal wastes can be handled and transported under more relaxed rules compared to hazardous wastes. However, e-wastes contain hazardous materials and must be taken to a designated handler or recycler. You may use this map to find an e-waste handler or recycler in your county, or visit the Cal-Recycler's database of the companies that collect, reuse and recycle electronic wastes.

The California Code of Regulations has developed a law under the title of social security, division: Environmental Health Standards for the

Management of Hazardous Waste

There are strict orders from the government for the proper implementation of rules in e-waste management that all e waste handlers (e.g. EDs, CRTs, and CRT glass handlers) who accumulate more than 5,000 kg of these UW at any one time will need to obtain a State generator ID number. Previously, CRTs and UWEDs were not counted towards the 5,000 kg that triggered the requirement to get an ID number.

Note: ***The International Network for Environmental Compliance and Enforcement*** (INECE) is a global network of environmental compliance and enforcement practitioners *dedicated to raising awareness* of *compliance and enforcement across the regulatory* cycle; developing networks for enforcement cooperation; and strengthening capacity to implement and enforce environmental requirements. Founded in 1989 by The Netherlands Ministry, of Housing; Spatial Planning and the Environment (VROM and by *the United States Environmental Protection* Agency. (US *EPA); INECE links the* environmental compliance and enforcement efforts of more than 4,000 practitioners inspectors, prosecutors, regulators, parliamentarians, judges, international organizations, and non-governmental organizations from 120 countries.

Initiatives and Measures Taken By Firms

Ways to Handle

In developed countries e-waste handling and processing involves dismantling the equipment into various parts like metal frames, power supplies, circuit boards, plastics, often by hand. This activity has its own advantages and disadvantages, where advantages are that the manual handling of e-waste can help in recognizing the useful components out of waste and those components which are in working mode or can be repaired including chips, transistors, RAM etc. On the other hand the disadvantages are that, the supply of labour is availed at very cheap prices and with lowest safety, and health standards. There following ways are generally appropriate in recycling/handling of e-waste:

1. Consumer Recycling

This defines the method of maximum utilization of any electronic device. Here the equipment is donated to organizations in need or the devices are directly sent to their original manufacturers, or supplying components to a convenient recycler or re-furbishers.

In this technique the systems or devices are not dumped as waste. Easily and are utilized optimally. There is variety of donation options, like charity that even offers taxi benefits.

Note: *The U.S. Environmental Protection Agency maintains a list of electronic recycling and donation options for American consumers. The National Cristina Foundation, Tech Soup (the Donate Hardware List), the Computer Take back Campaign and the National Technology Recycling Project* provide *resources for recycling. However the local recycling sites contribute to e-waste, when leaves the electronics trash thrown on sites unprocessed.*

2. Take back

This system is based on the social ethics that organizations must follow, to show their involvement and concern towards the society and to develop goodwill in the market, for this the companies are supposed to step forward to solve the ill issues within the society arising due to corporate interference.

Several companies have started working on this concept of *take back, in* which the consumer can return their electronic goods after use, to the company, to get it recycled. The category of such electronic goods includes mobile phones, computers, laptops, digital cameras and home and auto electronics.

Many well known brands have shown their participation in this, such as Staples, Gateway, Nokia, etc.; these companies provide monetary incentive for working technologies or those which can be at least recycled and can be used' again.

The "Panasonic Sharp Corporations & Toshiba founded "The Manufacturers Recycling Management Co." to manage electronic waste that are branded by the same companies, all such goods includes, 750 tons of TVs, Computers, Audio Equipment, Faxes.

Office depots let's customers obtain "Tech Recycling" boxes collection of e waste in case they are not eligible for the Eco NEW tech trade in program.

Hewlett Packard has recycled over 750 million pounds of electronic 'waste globally, including hardware and print cartridges.

The disadvantage of this program is that many corporations offer services for a variety of electronic items, while their recycling centres are few in numbers. This system although provided in several countries but the type and amount of equipment to be recycled tends to be limited. Sony participate in this take *back* programme but with a limitation that it takes only five items in a day and the product should be company's own production.

Planning for the eventual end of life management of electronic products generally takes a backseat to performance and price considerations when making purchasing decisions. Fortunately there are resources available to provide guidance on environmentally preferred procurement practices.

Nokia has designers who think carefully about the materials they use and they choose eco efficient recycling companies to take care of the recycling of the products from Nokia, 65=80% of the materials in a Nokia mobile phone can now be recycled and given a second life. Best practices can recover 100 percent of the materials partly as energy. Cover parts, of these phones are, clearly marked as recyclable.

Issues Related to Recyling of E-waste

1. It appears highly expensive and perhaps leads to raised disposal cost for the brokers and so-called recyclers, if they go for removal of items like CRTs (the procession of which is highly expensive and difficult). Thus they usually export unscreened electronic waste to developing countries to avoid such heavy expenses.
2. The defenders of the trade in used electronics also states that the Hard Rock mining of copper, silver, gold and other materials extracted from electronics is far more environmental damaging than the recycling of those materials. They also state that concept of refurbishing has

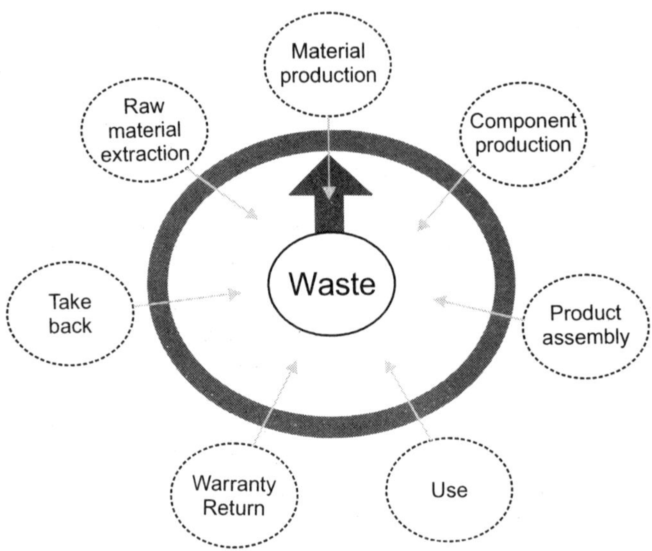

Figure: Represents the waste management technique adopted by Nokia

The Ad campaign by Nokia as "Planet ke Rakhwale*" and the punch line* "We Recycle*" is another major example of the initiatives that companies are taking to handle e-waste in best possible ways.*

nearly killed the art 3 of repair and reuse of electronic items. South Korea, Taiwan and Southern China have excelled in the art of refurbishing and now they own a set up of billion dollar industries in refurbishing items like Ink-Cartridges, Single Use Cameras and working CRTs.

Note: *Refurbishing has traditionally been a threat to established manufacturing, and simple protectionism explains some criticism of the trade. Works like 'The Waste Makers" by Vance Packard explain some of the criticism* of *exports* of *working product, for example the ban on import of tested working Pentium 4 laptops to China, or the bans on export of used surplus working electronics by Japan.*

3. The opponents of "Surplus Electronics Exports" argue that due to over environmental and labour standards, relatively high value of recovered raw material leads to transfer of pollution generating activities. For example, burning of copper wire.

4. There is an illegal supply of electronic waste in countries like China, Malaysia, India, Kenya and various African countries. Because the United States has not ratified the Basel Convention or its Ban .Amendment; and has no domestic laws forbidding the export o£ toxic waste; the Basel Action Network estimates that about 80% of the electronic waste directed to recycling in the U.S. does not get recycled there at all, but is put on container ships and sent to countries such as China. This figure is disputed as an exaggeration by the EPA, the Institute for Scrap Recycling Industries, and the World Reuse, Repair and Recycling Association.

   Note: *Guiyu in the Shantou region of China, Delhi and Bangalore in India as well as the Agbogbloshie site near Accra, Ghana have electronic waste processing area. Uncontrolled burning, disassembly, and disposal can cause a variety' of environmental problems such as ground water contamination, atmospheric pollution; or even water pollution either by immediate discharge or due to surface run off (especially near costal areas), as well as health problems including occupational safety and health effects among those directly involve, due to the methods of processing the waste. Thousands of men, women and children are employed in highly polluting, primitive recycling technologies, extracting the metals, toners, and plastics from computers and other electronic waste.*
5. The other major issue that is a matter of immense concern and attention is; several cities and slender tall buildings are getting developed on the hundreds of abandoned landfills cropping up due to the real estate boom. This may lead to the supply of contaminated water that can come from the soil nearby and thus can result in harmful effects to human life.
6. Another most important problem that has been observed is the resistance of several companies towards ; handling over, their computers to the organizations, involved: in recycling of such goods because they have threats of data leakage and its security.

Research and Analysis of Data

The research and analysis of the data represents the various aspects related to e-waste such as the probable sources of e-waste, countries producing e-waste and the percentage of e-waste produced by them, etc.; this analysis is a basic support to draw conclusion about the degrading environment due to increasing amount of e-waste also what measures should be taken by the government and the world E-waste Handlers to prevent further environmental degradation.

Major Generators of E-waste in India

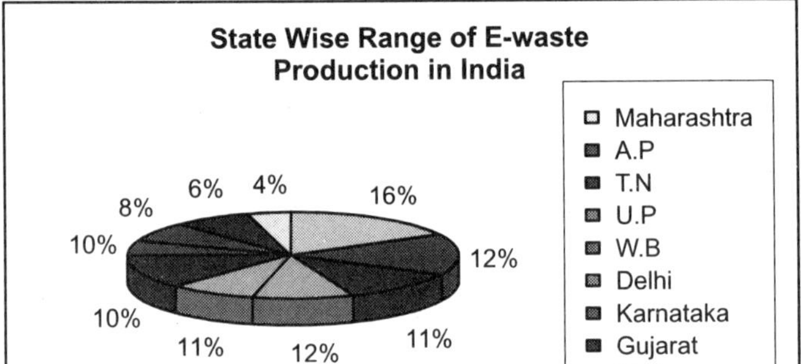

1. **Analysis:** *The following chart represents the ascending order of states in India that are major producers of e-waste.*

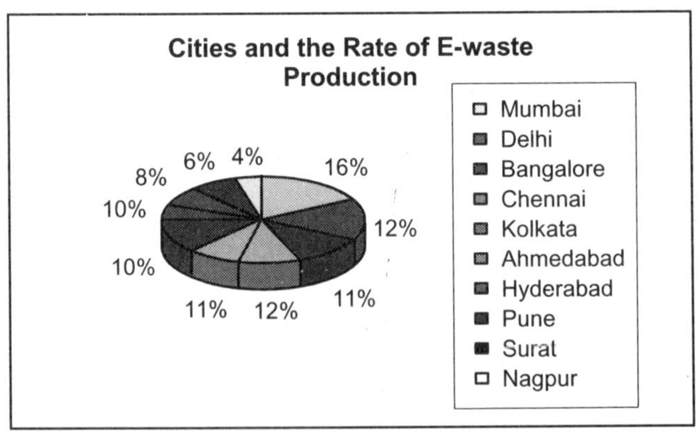

2. Analysis: *The figure shows the city-wise status of e-waste generation in India.*

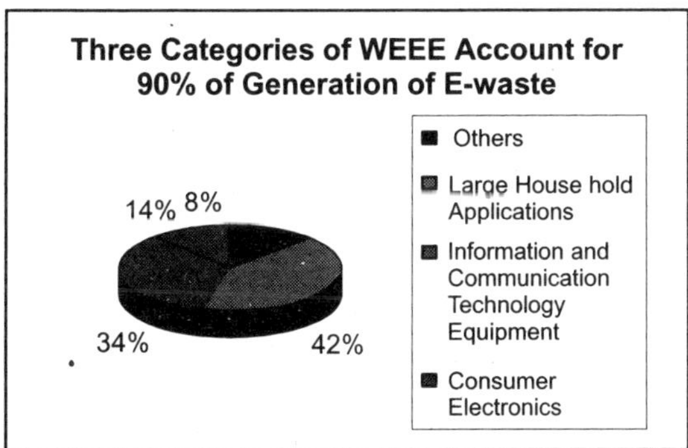

3. Analysis: The figure represents the categories and percentage of e-waste that comes under WEEE.

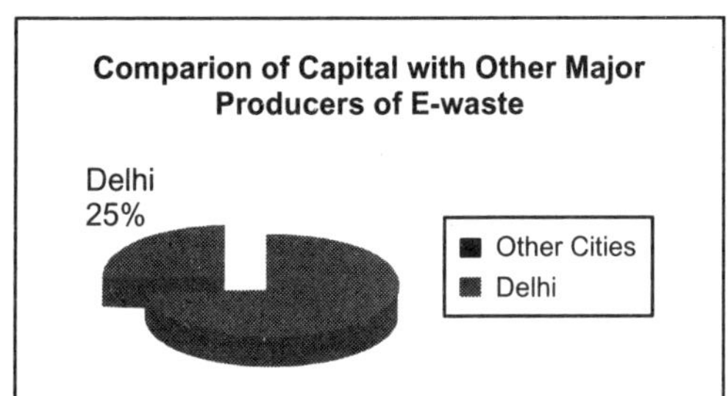

4. Analysis: *India produces nearly 146002 tones of e-waste a year, in which Delhi alone contributes to 25% of generation of e-waste.*

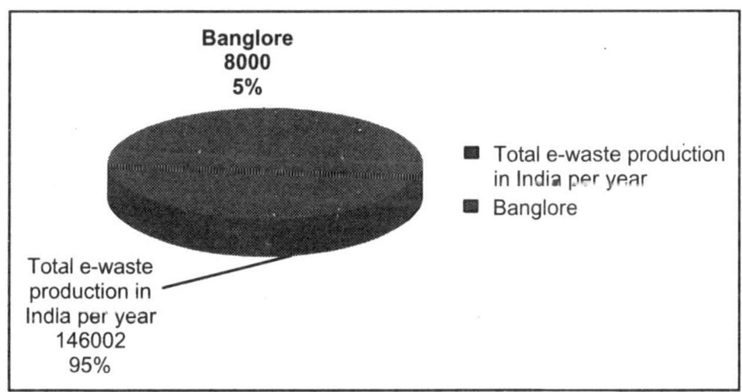

5. Analysis: The chart represents the share of e-waste production in Bangalore as compared to the entire country.

Conclusion and Suggestions

One cannot ignore the growth and development in technology that man has come up with, in the recent few decades. This development in science and technology has no doubt, made human life more comfortable and easier, but on the other side of this success there lies another major issue that is equally noticeable. The faster rate of growing technological changes will soon become a curse for mankind when there will be no scope left for managing the excess of e-waste that is increasing with increasing technological changes.

Though every country is facing this challenge of managing e-waste but the one which is faced by India is one of concern. In India the increasing volume of e-waste is not only domestically generated but the country imports e-waste from other developed countries even. Here the government should try to frame such policies so as to ban this system of import.

Owing to the illegal supply and transport of e-waste there is no exact estimate of the amount of e-waste that is supposed to be recycled.

The government and other universal waste-handlers should take it as a responsibility to increase awareness among the manufacturers for the hazards related to incorrect e-waste disposal.

There should be proper watch over the people generating business for dealers in e-waste and those who use rudimentary techniques like acid leaching and open air burning resulting in severe environmental damage. It is important to make them aware of the health hazards that can occur if they continue with such activities.

The use of computers as everyone knows is not confined just to the IT industry but is explored in almost every possible kind of industry as people are becoming more and more techno-friendly this is, probably leading towards more and more domestic production of e-waste. The companies should get associated with the local electronic recycling centre in their area to find the best possible way to manage their company's e-waste production. These centres are more or less working on the concept of Green Computing or Green IT, which is a study or practice of designing, manufacturing, using and disposing of electronic goods in a way so that it makes minimal or no difference to the environment. This study is even concern about the cost involved in recycling and disposal and is thus a practice of effective usage of resources.

It has become the need of the day to develop certain cohesive national policies to manage e-waste or else this may lead to further degradation of the planet and perhaps it will become an issue that will go out of control.

REFERENCES

Books/ Derivatives

Electronic Waste Management: *Design, Analysis and Application Series: Issues in Environmental Science and Technology, Vol. 27 by Hester, Ronald E.; Harrison, Roy, M. (Eds)*

E-waste: Managing the Digital Dump Yard, *by Vishakha Munshi*

Links/ websites

http://www.mumbaispace.com/issues/e waste.htm

http://www.nokia.co.in/about nokia/environment/we recycle/recycling-through -life-cycle

http://www.ccr.oal.ca.gov.

http://www.calrec cle.ca.gov/electronics/Collection/RecyclerSearch.aspx

http://www.youtube.com./watch?v =dIL3OWpz2no

http://www.naturalnews.com/e-waste.html

http://www.mait.com/admin/press images/press 16dec09.htm

http://www.sciencedirect.com/science? ob=ArticleURL& udi=B6V9G 4G9RIBN-2& user=10& coverDate=07%2F31%2F2005& rdoc=1& fmt=hi & oriw=search&sort=d& docanchor=&view c& searchStrId=1188357497& rerunOrigin=goog le& acct=000050221& version=1& urlVersion=0& userid=10&md5=66a0115cb6b9b999ba7d af689deb12a0

http://www. scribd. com/doc/20784043/Managing E Waste Indian Persoective.

http://www.business standard.com/india/news/law to make producers responsible for e-waste/3 79656/

6

Pollution in India

Kumud Dandotiya and Kamakshi Dandotiya

Air Pollution

Poor air quality is one of the most serious environmental problems in urban areas around the world, especially in developing countries. Recent studies that assess and value the adverse health impacts of exposure to particulates reveal the magnitude of the costs to society that calls for immediate actions. The paper shows that India appears to bear a very high level of these costs by international comparison. It reviews some latest findings in quantifying the impact of exposure to particulates on mortality with a special reference to India, and discusses the issues of economic valuation of sickness and premature death due to air pollution, with the focus on developing countries. Further, the paper analyses, drawing upon a case study of Mumbai, the relative effects of various pollution sources on the exposure levels and health outcomes, as well as the health benefits of specific control measures and policies. The concluding section highlights a set of issues and recommendations regarding a better integration of environmental health considerations into pollution management decisions.

Industrialization and urbanization have resulted in a profound deterioration of India's air quality. Of the 3 million premature deaths in the world that occur each year due to outdoor and indoor air pollution, the highest number are assessed to occur in India. According to the World Health Organization, the capital city of

New Delhi is one of the top ten most polluted cities in the world. Surveys indicate that in New Delhi the incidence of respiratory diseases due to air pollution is about 12 times the national average.

According to another study, while India's gross domestic product has increased 2.5 times over the past two decades, vehicular pollution has increased eight times, while pollution from industries has quadrupled. Sources of air pollution, India's most severe environmental problem, come in several forms, including vehicular emissions and untreated industrial smoke. Apart from rapid industrialization, urbanization has resulted in the emergence of industrial centres without a corresponding growth in civic amenities and pollution control mechanisms.

Regulatory reforms aimed at improving the air pollution problem in cities such as New Delhi have been quite difficult to implement, however. For example, India's Supreme Court recently lifted a ruling that it imposed two years ago which required all public transport vehicles in New Delhi to switch to compressed natural gas (CNG) engines by April 1, 2001. This ruling, however, led to the disappearance of some 15,000 taxis and 10,000 buses from the city, creating public protests, riots, and widespread "commuter chaos." The court was similarly unsuccessful in 2000, when it attempted to ban all public vehicles that were more than 15 years old and ordered the introduction of unleaded gasoline and CNG. India's high concentration of pollution is not due to a lack of effort in building a sound environmental legal regime, but rather to a lack of enforcement at the local level. Efforts are currently underway to change this as new specifications are being adopted for auto emissions, which currently account for approximately 70% of air pollution. In the absence of coordinated government efforts, including stricter enforcement, this figure is likely to rise in the coming years due to the sheer increase in vehicle ownership.

Waste and Water Pollution

Water pollution has many sources. The most polluting of them are the city sewage and industrial waste discharged into the rivers. The facilities to treat waste water are not adequate in any city in India. Presently, only about 10% of the waste water generated is treated;

the rest is discharged as it is into our water bodies. Due to this, pollutants enter groundwater, rivers, and other water, bodies. Such water, which ultimately ends up in our households, is often highly contaminated and carries disease causing microbes. Agricultural run off, or the water from the fields that drains into rivers, is another major water pollutant as it contains fertilizers and pesticides.

During the last fifty years, the number of industries in India has grown rapidly. But water pollution is concentrated within a few subsectors, mainly in the form of toxic wastes and organic pollutants. Out of this a large portion can be traced to the processing of industrial chemicals and to the food products industry. In fact, a number of large and medium sized industries in the region covered by the Ganga Action Plan do not have adequate effluent treatment facilities. Most of these defaulting industries are sugar mills, distilleries, leather processing industries, and thermal power stations. Most major industries have treatment facilities for industrial effluents. But this is not the case with small scale industries, which cannot afford enormous investments in pollution control equipment as their profit margin is very slender.

Chemical Pollution

As rapidly developing countries such as India industrialize, the dangers to local communities from pollution are often overlooked until there is a major disaster such as occurred in Bhopal.

The effects of chemical pollution are being rapidly felt across India. It has found that the incidence of diseases related to nervous, circulatory, respiratory, digestive and endocrine systems was one to four times higher in heavily industrialized areas as compared to unindustrialized areas. Many cases of congenital deformity and chromosomal abnormalities were also reported, in addition to 11 cases of different kinds of cancer. Skin disorders are also rampant.

The wave of industrialization that began in the late 1970s has changed the complexion of India's once placid landscape. Lakes, streams, as well as the groundwater are laced with toxic heavy metals and chemicals, as proved by several studies by government agencies and research institutions including the National Geophysical Research Laboratory.

7

The Impact of Human Activities on Environment

Savita Shrivastav, Mukesh Sharma and P. Chaudhry

Since the existence of life on earth man has been getting good favours to enjoy health and luxurious living from environment. There is a close relationship between man and environment. We the human and other living beings, are affected by the changes which happen in the atmosphere and environment. But we also change and affect the various normal and balanced conditions of environment. We have the right to make progress and development in every field of life and society. It is made possible through the various rich and life giving resources available in a natural environment. In the ancient times, when there was no problem from population, human beings used nature and natural resources for their needs, but due to limited requirement the use of the natural resources was done in a controlled way. In the course of development in every field of life there came a large increase in population. As the resources available on earth are limited, the rapid flow of population is bound to misuse the natural resources in an uncontrolled way. The needs of the progressive society are rising and nature has to provide for them from its limited sources. It is obvious to become problematic for the well wishers of man in relation to the natural environment. Moreover to keep pace with the fast running modern and advanced society men care only for their selfish ends and they do not take heed of the harm and damage being done by their activities to the climate, atmosphere and environment.

The impact of human activities on environment can be studied at many levels. Human beings have affected and are still changing the healthy living conditions on earth in many ways. The different processes and systems involved in the course of development have drawn the attention of the scientists of the world. Life on earth is becoming difficult because of different changes in environment. In this paper I shall concentrate on environmental degradation, pollution, deforestation and desertification. Moreover a brief look at the impact of human activities on climate vegetation, water, animals and soil is also necessary to be mentioned here. The central and state governments of India have been framing a number of acts, laws and constitutional provisions to protect and conserve environment, but the acts and regulations will not merely work in this direction. Human beings on earth have every right to enjoy life the way they like but it is necessary to keep the house clean and worth living in which we are and we aspire to live in good health. One more important factor needs our attention in this direction. We will have to abridge the wide gap between the enjoyment of our rights and the performance of our duties. We expect more but do not give more in turn.

The Impact of Human Activities on Climate and Atmosphere

The two main factors in increasing human population and rise of technology are responsible for the different changes in atmospheric conditions as well as pollution. The climate of a particular region is changed due to gas emissions, thermal pollution, deforestation, overgrazing, extension of irrigation and diversion of fresh waters into oceans. One other major cause of change in atmosphere is the air pollution. It gives rise to the global warming and is caused mainly by the emission of carbon dioxide (CO_2). The large scale misuse of the stored carbon out of the earth in the form of coal, petroleum and natural gas disturbs the natural balance of environment. These things are burnt by man in industries which results in the emission of carbon dioxide (CO_2) and sulpher dioxide (SO_2). The emission of these two harmful gasses is increasing air pollution, surface temperature, green house effect and ultimately the global warming. In this way the impact of human activities on atmosphere is more harmful for the atmosphere work as a medium of transfer of the pollutants from one place to another at long distances and a major part of environment is disturbed.

Impact on Vegetation and Animals

The impact of human activities on vegetation and animals is worth notice. Human beings in the race of development have been in need of more and more place and land for habitation and cultivation. Here, the clearance of forests by fire or removal of trees and plants has become harmful for vegetation and animals. The uncontrolled burning and felling of woods with heavy grazing is responsible for the disappearance of vegetation cover and extension of wildlife. The increase in arid and semi-arid areas and desert land is all due to human activities. The harmful air pollutants, like SO_2 released by man into the atmosphere and the industrial fumes have killed many plants and have ruined vegetation.

Most of the developed and under developed countries are facing the ecological problem of extinction of several rare animal species. The extinction is partly due to excessive intentional killing for subsistence and commercial purposes but mainly due to pollution, change in ecosystems and natural habitats. We know that the presence of heavy metals and methyl mercury in water is harmful for the creatures living in it. The oil pollution is dangerous for marine and coastal fauna and flora. The feathers of the sea birds are clogged owing to oil pollution in water.

Impact on Soil and Waters

It is a universal fact that the life of the living beings including man is close to and depends on soil. In the course of development human activities have changed the very nature and structure of soil. The character of soil has been changed by agricultural machinery, irrigation, grazing through trampling and by the excessive use of chemicals to increase the productivity of agriculture. The use of such chemical fertilizers like nitrates, phosphates and potash has certainly increased production in agriculture but it has also changed the basic structure of soil. The nature and character of soil is also being changed constantly by large scale construction, urbanization and mining.

Water is the most essential and life giving source for man and is used for various basic and urgent needs on earth. The safe future of the increasing population and modern developments depends on pure water. In the course of progress human activities have influenced both the quality and the quantity of water. The construction of dams and reservoirs are useful for generating power and to provide a reliable source of water. But its impact can be seen in the changing ecosystem and environmental conditions. The negative effects of dams and reservoirs are subsidence, earthquakes, transmission and expansion in the range of organisms, changes in ground water levels, water logging and deforestation. Urbanization has also been a great cause of water pollution. Construction of sewerage lines and septic tanks, pollution of streams and wells, drilling of deeper, large capacity and industrial wells and excessive use of water for domestic and commercial purposes have changed the conditions for healthy living in the polluted environment.

Factors Responsible for the Deterioration of Environment

Man has been making progress since the beginning of civilization. It needs much extraction of natural resources. The development process and the extraction of natural resources has filled the environment with the disposal of waste material. We know the main things that have caused the environmental deterioration. First and foremost is the density and growth of population. It has increased pressure on natural resources. It is also responsible for severe exploitation and environmental degradation. The second thing is the developed technology. The fuel-based modern technology has made the large scale production possible but at the same time it has exerted a great strain on natural environment. The third major factor responsible for the degradation and deterioration of environment is industrial development. The above mentioned three factors are the creation of human beings for their benefit and needs.

Impact of Development on Environment

Man is a social being and his socio-economic development depends solely on environment. In the ancient times man regarded environment as the composition of the elements of nature. Then they considered it as natural environment. There are two main segments of natural environment, the physical and the biospheric. Here the biosphere depends on the physical for its survival. Man with the development in his capacity has utilized the natural environment to fulfil his basic needs and desires. The developed man improved modem technology which resulted in the large scale exploitation of natural resources. Thus man became the destroyer of nature by degrading and polluting the natural environment through his developmental activities.

Deforestation and Desertification

The destruction and clearance of forests, trees and plants causes deforestation. Its result is desertification which is caused mainly by overgrazing indiscriminate felling of trees and over exploitation of land resources. The growing demand of cities and the dire need for industrial areas are equally responsible for deforestation. The large scale deforestation is adversely affecting the weather and climate of different regions of the world.

Desertification is the result partly of the change in climate and partly of the abusive way of land use. Humans have been removing the vegetal cover of fertile land and it brings changes in the climate of a particular area. Desertification and deforestation are bringing major changes in rainfall, temperature and wind velocity. It is no denying the fact that man has done considerable damage to forests and the different ecosystems. According to sources, "At the time of independence in the country 75 million hectares about 22% was under forest cover. Today this has reduced to just 10%. India has been losing 10 million trees every year.

Constitutional Provisions and Mass Awareness

It has universally been acknowledged that air, water, land, animals and human beings are the creation of some unknown but superior power. Therefore it is our first duty to live in harmony with each other as each one of these is dependent on one another and damage or destruction to any of these means the same loss to all things. Our ancient holy scriptures gave the message to live in harmony with nature. After and before independence there have been many laws framed to protect environment. The constitution and the government of India have made many acts and laws for the protection of conservation of environment. Environment (Protection) Act 1986, Air (Protection and Control of Pollution) Act 1981, Water (Prevention and Control of Pollution) Act 1974, Noise Pollution (Regulation and Control) Rules 2000, Wildlife (Protection) Act 1973, Hazardous Wastes Management Rules 1989, Bio-medical Wastes (Management and Handling) Rules 1989 and the National Environment Tribunal Act 1995, have been made to the safe protection and conservation of Environment. In addition to the above the Government of India has initiated and started plantation to maintain balance in the natural environment. But it is a common tendency to make use of any thing up to any extent for one's own profit and selfish ends. Until there is a mass awareness towards pollution and degradation in environment no law or provision will work properly in the saving of environment. So with the plans of development industrialization and urbanization we will have to fix accountability and responsibility at the individual level to watch work and consider for the protection of environment and for the healthy conditions for our life.

REFERENCES

1. Awasthi, H.M. and P.R.Pande. *Environmental Studies*, Educational Publishers, Agra.
2. Mishra, D.K. (ed.). *Population, Environment and Development*, A.P.H. Publishing Company, New Delhi, 2004.
4. Saxena, H.M.. *Environmental Geography*, Rawat Publications, Jaipur, 2007.
4. Sharma, P.D. *Ecology and Environment*, Rustogi Publications, Meerut, 2009.

8

Biopesticides: A True Alternative Solution of Pesticide Residue Poisoning

R.K. Jain, Archana Shrivastav and D.K. Sharma

Introduction

The annual loss due to the pest problem to crops has estimated about 25,000 crores, WHO estimates more than 20,000 death and 100,000 pesticide poisoning cases every year, globally, this is due to high level of pesticide residue in food and food products including cereals, pulses, vegetables, fruit, fish, poultry, meat products, milk, milk products, water and beverages (fruit syrups and cold drinks). The farmers are heavily dependent on chemical pesticides in the last three decades. Indiscriminate use of chemical pesticide has led to a serious environmental problem. The excessive and indiscriminate use of chemical pesticide has also led to development of resistance in many pests. The chemical pesticides causes some severe health hazards to human and other living organisms.

Considering the health hazards, environmental damages caused by chemical pesticides and the development of resistance by an increasing number of pests. There was an urgent need for the use of alternative methods of crop protection for a very long time. The potential use of microbial and other biopesticides for this purpose is one of the most appropriate and promising of these methods.

Biopesticides are certain types of pesticides that derived from such natural materials an animal, plant or microbial origin. Use of botanical material, bacteria, fungi, nematode, virus or other microbes studied as biopesticides and these processes of control of plant pathogen are called biological control.

Biological control of soil-borne pathogens was not considered commercially feasible about two decades ago. But currently the opinion has shifted and several companies are on the verge of developing biocontrol agents as commercial products. This shift in the important role of biocontrol agents in agriculture is mainly perhaps in response to the public opinion regarding the hazards of chemical pesticides.

Biological control of plant and crop diseases is increasing all over the world. In India biological control studies have been carried out at various centres such as Indian Agriculture Research Institute (IARI) New Delhi, Tamil Nadu Agriculture University (TNAU) and Project Director of Biological Control (PDBC) Banglore, etc. Many companies in the world produce such microbes to use them as biopesticides.

Why Biopesticides?

Chemical pesticides cause some severe health hazards to mankind including convulsions, carcinogenicity, leukemia, neurotoxicity, dermatitis, allergies, paralysis to human health and environmental pollution as well. Beyond these health hazardous chemical pesticides have the following disadvantages:

1. Ozone depletion
2. Carcinogenicity
3. Mutagenicity
4. Ground water pollution
5. Air pollution
6. Acute toxicity (to worker)
7. Harm to non-target organisms and endangered species
8. Residue remains in food
9. Pests develop resistance to chemical pesticides.

Beyond harmful effects of chemical pesticides the biopesticides have the following advantages:

1. Biopesticide can be used for prevention of plant pathogens including bacteria, fungi, nematode or other kinds of pathogens.
2. Microbial biopesticides are odourless and do not leave harmful residue in crops.
3. Non-toxic to human, animal and plants.
4. All kinds of biopesticide are target specific and biodegradable thus they do not destroy beneficial organisms.
5. They promote the growth of natural enemies of pests thus reducing the need for future pesticide application.
6. They remain effective at all stages of microbial pathogens.
7. Being environmental-friendly does not disturb the ecological balance.
8. They are also effective against those pathogens that have resistance to chemical pesticides.

3. Kinds of Biopesticides

Plant pathogen attacked by a multitude of microbes. A number of these are adaptable for mechanical dissemination as microbial pesticides. At present, botanical material, bacteria, fungi, viruses and nematodes have been introduced for the successful control of Insects, fungus or other kinds of plant pathogens. Some of the microbial pesticides are described as under:

(A) Botanical Materials as Biopesticide

There are so many plants and plant products that can be used as biopesticide including Neem (Azdirectin indica), Karanj, Babool and Echinesia echinatum. Neem oil and Azadirectin based products of 300 ppm, 1500 ppm and 10000 ppm are available in the world and Indian market with good potential.

Distillation product of neem leaves, karanj leaves and Echinesia root are low cost biofungicide can be produced by farmers at their home.

(B) Bacteria as Biopesticide

There is a wide range of bacteria that control bacterial, fungal, nematode and insect pathogens described under:

1. **Pseudomonas fluorescence:** It controls Erwania emylovera in apple, cherry, almond, peach, pear, potato, strawberry and tomato, etc. it also works against so many other soil born fungal pathogens including Rhizoctonia of root rot of cotton. Nowadays this biopesticide has a very good market potential.
2. **Bacillus thuringiensis:** They produce a protein crystal named 0 exotoxine that causes paralysis in caterpillars and their larvae and non-toxic to other life forms including human, animal and plants. It is very effective against caterpillars and their larvae of Cotton, Alfa alfa, Beans, Cabbage, Cucumber, Cauliflower, Sweet Corn, Tobacco, Tomatoes, Oranges and Grapes etc.
3. **Bacillus subtilis:** It is a gram positive bacteria that control a wide range of fungal plant pathogens including Rhizoctonia, Fusarium, Alternaria and Aspergillus that cause root rot and seedling diseases. It is very effective against scab, powdery mildew, downy mildew, sour root, early leaf spot, early blight, late blight, bacterial spot and walnut blight disease of plants.
4. **Streptomyces Sp.:** It is used to control mites and spider mites of agriculture crops and collardo beetles of potato.

(C) Fungi as Biopesticide

Fungi have received most attention as antagonist, possibly because of the ease in handling and identification as compared to other microbes. Some of the fungi uses as biopesticide are described as under:

I. Fungi as Biofungiicide:

1. **Trichoderma viridi:** It is used to control some important soil born fungal pathogens and control most of the fungal diseases of plant roots, stems, leaves and seeds including Loose smut of wheat, black pod, Frosty pod rot and witches (brooms) of cacao, Fusarium wilt of muskmelon, etc. It is the most important Biofungicide having a very good market potential.
2. **Gliocladium virens:** It also works against soil born fungal

pathogens including Rhizoctonia, Pythium and Sclerotium that causes coller rot and root rot in Beans and ornamentals.

II. Fungi as Biopestcide or Bioinsecticide:

1. **Beuveria bassians:** It works mostly against hairy caterpillars including insects, eggs of Lepidopteron, Moth, Chiggers, White grubs, Fire ants, Flea beetles, White flies, Plant bugs, Grasshoppers, Thrips, Aphids, Mites, etc. It is one of the most aboundent Bioinsecticides.
2. **Metarrhizium anisopliae:** It works mostly against Arthopodes including grasshoppers and soil inhabiting beetles and termites and other harmful insects. It is also one of the most abundant bioinsecticides.
3. **Verticillum lecanii:** It works mostly against sucking pests including whiteflies and aphids and very effective against aphid having a very good market potential.
4. **Paicilomyces fumosorosus:** It controls whiteflies, Aphids and Thrips.

II. Fungi as Bionematicide:

1. **Paicilomyces lilacinusus:** It controls root knot nematodes and other nematodes.
2. **Dectylella oviparacitica:** It is also found active against nematodes.

(D) Viruse as Biopesticide

Viruses are obligatory parasites, Viruses have successfully grown in Vitro but their large scale production is difficult and an invasive method. Some of the viral pesticides are described below:

1. **Nuclear Polyhydrosis Viruse** (NPV): Nuclear Polyhydrosis Viruses are host specific nemetopathogens. They multiply only in specific host. Thus their mass production is undoubtedly linked to the mass production of the specific host cells that's caterpillars. Thus a few kinds of NPV are possible including

(*i*) **Heliothis NPV:** This viruse grows on Heliothis armigera caterpillar host and effective against caterpillars of cotton, maize, tomato, groundnut, sunflower and vegetables.

(*ii*) **Spodoptera NPV:** This viruse grows on Spodoptera litura caterpillar host and effective against caterpillars of cotton, tobacco, castor, pulses and oil seeds.

2. **Chilo infuscatellus granulosis viruses** (GV): This belongs to granulosis viruses similar to other baculoviruses and is a very target specific attack only on Chilo infuscatellus (sugarcane borer), which is an important pest of sugarcane.

3. **Trichogramma spp:** for control specific lepidopteron caterpillars.

(E) Insect and Other Organism as Biopesticide

1. Epiricania : for control of Pyrilla in sugarcane plant
2. Chrysoperla camea: for control of aphid, jassids, whiteflies and caterpillars of agricultural crops.

4. Product Description or Formulation

To be a desirable microbial pesticide, the agent should not diminish in potency by incorporate processing. The choice of formulation may be vitally important in practical application. For example, if heavy rainfall is expected, dust would be unsuitable and a spray containing a good sticker would be required.

There are two main types of formulation of microbial pesticides

a. Liquid Formulations
b. Dust Formulations

Additives, namely spreaders, stickers and protectants are used in the formulations for wetting, to form a weather resistant film on foliage and to minimize the effects of desiccation, sunlight and/or UV radiation.

Usually all the bacterial preparations may be prepared as either liquid formulation or as a carrier-based formulation. The fungal preparation is prepared in carrier-based formulation but there are possibilities to prepare liquid formulation of fungal biopesticides however viral formulation is prepared only through liquid formulation.

5. Process Description

(a) Process for Trichoderma and Other Fungal Biopesticides

The trichoderma is grown in a suitable liquid culture medium in fermentors or in bottles, under controlled culture conditions. When the growth is complete the broth is collected and homogenized. At this stage, the spore load is adjusted to the required level. The suspension is either used as such or turned into powder formulation. For this the broth is mixed thoroughly with a suitable filler (carrier) at a suitable concentration. This is then dried and stored.

Procurement of Strain
→ Check Purity and Biochemical Testing
↓
Subculturing
↓
Preparation of Small Flask Culture
↓
Preparation of Large Flask Culture
↓
Harvesting of Broth After Completion of Growth
↓
Homoginization
→ Check Quality of Broth
↓
Suitable Fungal Concentration by Mixing in a Filler
→ Check Quality of Product
↓
Packing
↓
Dispatch for Sale

(b) Process for Bacillus and Other Bacterial Biopesticides

All the bacterial biopesticides prepared by growing them in liquid medium and after optimum growth occurs the broth is harvested in aseptic conditions and their liquid or dusty formulations are prepared by adding or blending them in a suitable filler or carrier.

Procurement of Strain

→ Check Purity and Biochemical Testing

↓

Subculturing

↓

Preparation of Small Flask Culture

↓

Preparation of Large Flask Culture

↓

Harvesting of Broth After Completion of Growth

→ Check Quality of Broth

↓

Mixing of the Broth in Liquid or Poder Carrier

→ Check Quality of Product

↓

Packing in Bags or Ldpe Bottles

↓

Dispatch for Sale

6. Preliminary Project Economics for Trichoderma

I. Net fixed cost = Rs. 25.5 lakhs

II. Plant running cost= Rs. 13.65 lakhs

Net profit of unit per annum at 100% capacity utilization = Rs. 31.76 lakhs per annum

I. Net fixed cost = Rs. 25.5 lakhs

a. Cost of Biopesticide facility for 50.0 to 100.0 M.T. per year = 11.6 lakh machines

S.No.	*Equipment*	*Quantity*	*Amount (Rs.)*
1.	Self steam generating Big Size Horizontal Autoclave (3×3×5 ft)	1	4,00,000
2.	Laminar air flow hood	1	1,00,000
3.	Microscope	1	1,00,000
4.	Big shaker machine (7×7 ft.) Able to rotate 39 No. 5.0 Ltr flasks	1	1,00,000
5.	Window A.C.	2	1,00,000

6.	Small shaker machine (3×3 ft.)	1	50,000
7.	Oven with hot air circulation	2	50,000
8.	BOD Incubator 6.1 Cu. ft.	1	50,000
9.	Top Loading Electronic Balance	1	25,000
10.	Laboratory glassware	Lump sum	75,000
11.	Mixing versal	1	75,000
12.	Automatic sealing	2	30,000
13.	PH Meter	1	15,000
14.	Refrigerator	1	15,000
	Total		11,60,000

b. Land = 2000 sq feet

c. Cost of building = Rs. 12.0 lakhs

d. Cost of electrical fixture, furniture and other miscellaneous items = Rs. 1.0 lakh

e. Fees for license for trichoderma production = Rs. 0.75 lakhs

f. Purchase of new effective strains = 0.15 lakhs

The total fixed cost from the above = 25.5 lakhs

Less subsidy on installation of Biopesticide plant = not known

Net fixed cost from the above = Rs. 25.5 lakhs

Interest @ 18% for 50.0 MT = Rs. 4.59 lakhs

Interest per MT= Rs. 9180

II. Plant Running Cost = Rs. 13.65 Lakhs

a. Establishment Cost:

S.No.	*Designation*	*No. of Employees*	*Annual Salary (Rs.)*
1.	Technical manager/ advisor	1	1,50,000
2.	Microbiologist/ Biotechnologist	1	1,20,000
3.	Lab technicians	1	75,000
4.	Peon-cum-lab attendant	1	40,000
5.	Daily rated labourers	As required	1,00,000
	Total =		4,85,000

b. Production Cost: (for 50.0 MT per year)

Chemicals and raw material Rs. 3,00,000 per annum (lump sum)

Energy charges	Rs. 1,50,000 lump sum per annum
Packing material	Rs. 2,00,000 lump sum per annum
Repair and maintenance	Rs. 30,000 lump sum per annum
Statutory obligations	Rs. 50,000 lump sum per annum

(Including Insurance and License renewal)

Marketing and transportation cost Rs. 1,50,000 lump sum per annum

Total running expenses per annum Rs. 13,65,000 per annum

Total running expenses =13,65,000/50 = Rs. 27,300/MT

Total expenses (Running + Fixed) for one MT

= Rs. 27,300+9,180= 36,480

Total expenses for one kg is = Rs 36.48

The sale price is Rs. 100.0/kg

Net profit = 100 36.48 = 63.52 per Kg = Rs. 6352 per M.T.

Net profit of unit per annum at 100% capacity utilization = 6352 × 50 = Rs. 31.76 Lakhs per annum

Note: This sale cost is only for Trichoderma. The other Fungal and Bacterial Biopesticides may also prepare at the same Production Cost, however they have a higher market price than Trichoderma. Thus the net profit is more than the estimated.

7. How Biopesticide Adoption Can Be Improved

1. State and central government should promote national Integrated Pest Management (IPM)
2. There should be a Committee of subject matter specialists for monitoring and observation of the effects of biopesticides, method of application, etc.
3. Central government should simplify the registration procedure of Biopesticides.

4. Government should promote organic food export by relaxation in the taxes and norms of export.

8. Summary

Pests have developed resistance against most of the chemical pesticides. Thus the chemical pesticides remain ineffective against pest control and also leave harmful pesticide residue in crops. Thus use of microbial biopesticide is increasing every day and nowadays so many companies are involved in production of microbial biopesticide commercially.

Microbial Biopesticide may be bacteria, fungi, viruses or even nematode origin; however, use of fungal and bacterial biopesticide is more common. Among series of bacteria Pseudomonas fluorescens, Bacillus subtilis and Bacillus thruingiensis are abundant bacterial biopesticide. Trichoderma, Beuveria, Metarrhizum and verticilum are commonly used Fungal Biopesticides. However Viral Biopesticides are difficult to produce commercially because of their invasiveness and host specificity.

All kinds of bacterial and fungal biopesticides are produced by growing them in a suitable liquid medium and after attending the abundant growth the broth is harvested and homogenized finally mixed in a suitable filler or carrier.

To establish a biopesticide industry is a matter of profit because for 50.0 MT Trichoderma Biopesticide the fix cost is about 25.5 lakhs, the plant running cost is about 13.65 lakhs and the profit is about 31.76 lakhs on 100% utilization of plant capacity. The government should promote some programmes for the biopesticide adoption improvement.

9

Impact of Humans on the Environment and the Implication of These Effects for Living Organisms

Manoj Sharma, R.S. Rathore and Arvind Dohre

Introduction

Pollution of our environment has become a serious problem and the study of this aspect has assumed unusual significance in ecology. Constitute the human environment pollution means direct or indirect change in one or more components of the ecosystems, which are harmful to the system or at least undesirable for man. There are several kinds of pollution and the causes are also many. There are natural mechanisms of homeostasis through which the ecosystem self regulates and maintains the ecological balance, in the process of exploiting natural resources. Man did not understand the significance of ecological balance. So instead of adjusting his actions according to ecological principles, present day humans through their technological activities have caused deterioration in their environment. The pace of environmental degradation is so rapid that the very survival of man may be at stake in the not too distant future. Different aspects of pollution are described by Ambashat and Ambashat (2005).

Sources of Environmental Pollution

In this admixture of gases constituting the natural atmosphere there are a large number of other substances that keep exchanging between the atmosphere and natural and man made sources.

1. Carbon Mono Oxide

Carbon mono oxide gas is formed by incomplete combustion of fossil fuel like oil and petroleum. Automobiles are the commonest source of carbon mono oxide pollution in the cities. Other sources are oil refineries, metallurgical operation and other internal combustion engines. It is a major pollutant for man and living organisms.

2. Carbon Dioxide

Carbon dioxide is a resource at its natural label of 0.03% as it is the raw material of photosynthetic reaction by which food is prepared by green plants and on this food ultimately all organisms including man depend. Carbon dioxide is a product of respiration from all living cells respire and burning of fossil fuel.

3. Chlorofluoro Carbon (C.F.C.)

Chlorofluoro carbons are chemicals synthesized by man for use in several kinds of industries including refrigeration and air conditioning and find their way into the atmosphere

4. Methane (CH_4)

Methane is another green house gas which is naturally produced. In recent years its production and release into the atmosphere appear to have increased due to the human influence. The main source are biological process such as enteric in cattle, sheep and other animals, an aerobic situation in wet lands and rice fields and burning of biomass and fossil fuel by human.

5. Nitrous Oxide (N_2O) and Sulphur Dioxide SO_2)

Nitrous oxide emitted from the burning of fossil fuel is another green house gas. Oxides of sulphur are the common pollutant being formed by burning of sulphur containing coal and petroleum since the quantity of burning of such fossil fuel is fantastic.

6. Ultra Violet Radiation

The main source of Ultra violet radiation is the sun. Ozone layer is reported to be getting thinner due to its slow breakdown by a number of chemical radicals produced by anthropogenic activities which move slowly upwards and reach the ozone shield. The most serious culprits are chlorofluro carbons.

7. Radioactive Pollution

Ionizing radiation are other very important and harmful pollutants. Nuclear war materials and test explosion are principal sources of radioactive wastes in the atmosphere.

8. Noise Pollution

Noise pollution is unlike all other pollution causing components of environment. The main sources are loudspeakers, fireworks and industries, vehicles, high sound crackers are worst offenders and aeroplanes.

Impact of Environmental Pollution

Carbon monoxide in man and animals as it combines with blood haemoglobin 200 times faster than oxygen does and produces suffocation. In nature the rise in CO_2 is necessarily accompanied with decrease in O_2 content which has harmful effects on human health. Thus CO_2 increases manifold adverse effects particularly all oxygen deficiency and green house effects on global weather and climate. Chloro fluoro carbon damages the stratospheric ozone layer. It is difficult to project the future picture for methane and its impact on green house effect or warming up of the global environment.

Ultra violet B is most dangerous to plants and animals. It creates cancer by mutation in genetic materials and depletion of good stratospheric ozone umbrella is responsible for generation of bad tropospheric through the enhanced ultraviolet B. The global agriculture will be adversely affected.

Ionizing radiation causes mutation, abnormality and lethality in many living organisms and human. Cancer is commonly caused even by unclear low label exposure. The high pressure sound can causes damage in a number of ways on our hearing ability, brain and balancing mechanism and excessive anxiety, fatigue, fright and change in heartbeat rates, dilation of pupil of eyes. Constriction of blood vessels, emotional disorders are more dangerous consequences of noise pollution.

Results and Dicussion

Results of researches on this aspects have been reviewed by Teramura (1983). It adversely affects plant physiological process and primary productivity. Ambasht (2005) et al. have made excellent reviews of the UV B impact of agriculture crops. In rice plant on UV B exposure the net production fell by 13 to 17%. (Ambasht and Agrawal, 1997)

Koromondy (1978) has regarded that fission by products bath of nuclear detonation and water cooled atomic power reaction do indeed constitute more of a potential hazard then direct ionizing radiation because they follow biochemical path ways. Singal (2000) has described noise pollution in respect of some Indian cities. He has complete data from different sources for the marketplace. The noise label reported in Delhi was between 86 102 dB, Kanpur 89 98dB and Lucknow 67 99dB.

Management of Environmental Pollution

1. Generating awareness about the causes of pollution so that preventive measures may be taken.
2. Monitoring the level of atmospheric pollution at weather stations in all principal cities and industrial areas.
3. Limiting the level of pollutant discharge.
4. Application of suitable antipollution measured by industries.
5. Management of aquatic bodies.

6. Construction of treatment tanks for treatment of organic wastage.
7. Restricting the use of non-biodegradable chemicals.
8. Use of such plants can be to harvest out the pollutants in order to keep the environment clean.
9. In Indian cities noise pollution control norms and low enforces.
10. Research effect to locate and research pollution sources with safe needs.

REFERENCES

Ambasht, N.K. and Ambasht, R.S. (2005). Plant Responses to Changing Ozone, and UV B Scenario. A Review. Proc. Nat. Acad. Sci. India, 75:159-168.

Ambasht, N.K. and Agrawal, M. (1997). Influence of Sapplemental UV B Radiation on Photo Synthetic Characterstics of Rice Plants, Photo Synthetatic, 34: 401-408.

Koromondy, E.J. (1996).

Singhal, S.P. (2000) Noise Pollution in Indian Cities in Environment Hazard Planet People (Eds Iqwal) Distributors New Delhi.

Teramusa, (1983). Effect of Ultraviolet-B Radiation on the Growth and Yield of Crop Plants Physio Plant 58: 415-427.

10

Bio-Medical Waste (Management and Handling)

A.R. Khan, Shema Khan and Sushil Gupta

Bio-Medical Waste

Waste generated in the diagnosis, treatment and immunization of human beings or animals, in research or in the production and testing of biological products is called bio-medical waste. Bio-medical waste consists of solids, liquids, sharps, and laboratory waste that are potentially infectious or dangerous. It must be properly managed to protect the general public, specifically healthcare and sanitation workers who are regularly exposed to biomedical waste as an occupational hazard.

Bio-medical waste differs from other types of hazardous waste, such as industrial waste, in that it comes from biological sources or is used in the diagnosis, prevention, or treatment of diseases. Common producers of biomedical waste include hospitals health clinics, nursing homes, medical research laboratories, offices of physicians, dentists, and veterinarians, home health care, and funeral homes.

A hazardous waste is waste that poses substantial or potential threats to public health or the environment and generally exhibits one or more of these characteristics:

- carcinogenic
- ignitable (i.e. *flammable*)
- oxidant
- corrosive
- toxic
- radioactive
- explosive

U.S. environmental laws (*Resource Conservation and Recover; Act*) additionally describe a "hazardous waste" as a waste (usually a solid waste) that has the potential to:

- cause or significantly contribute to an increase in mortality (death) or an increase in serious irreversible, or incapacitating reversible illness; or
- pose a substantial (present or potential) hazard to human health or the environment when improperly treated, stored, transported or disposed of or otherwise managed.

These wastes may be found in different physical states such as gases, liquids, or solids. Furthermore, a hazardous waste is a special type of waste because it cannot be disposed of by common means like other byproducts of our everyday life. Depending on the physical state of the waste, treatment and solidification processes might be available. In other cases, however, there is not much that can be done to prevent harm.

Medical waste, also known as clinical waste, normally refers to waste products that cannot be considered general waste, produced from health care premises, such as hospitals, clinics, doctors' offices, labs and nursing homes.

Problems of Bio-Medical Waste

The perils of medical *waste* first drew attention in the late 1980s, when items such as used syringes washed up on several east coast beaches of the USA which lead to the law regulating the medical **waste**. However in India the seriousness about the **management** came into limelight only after the 1990s.

"**Bio-medical waste**" means any waste, which is generated during the diagnosis, treatment or immunization of human beings or animals or in research activities pettaining thereto or in the production or testing of biological, and including categories mentioned in Schedule 1.

Medical wastes may be generated throughout health care facilities wherever medical procedures are conducted. The ill effects of poor **management** of **bio-medical waste** have aroused concern all over the world especially due to its far reaching effects on human health and environment. Moreover, new hospitals are coming up at a mushrooming speed to meet the health hazards. Main cause for increase in quantity of medical **waste** is the shift from reusable items to disposables goods. The problem of medical **waste** has acquired gargantuan proportions and complex dimensions. On an average, a hospital bed generates 1 kg of **waste** per day, out of which 10 15 per cent is infectious and the rest is general **waste.**

The quantities become unmanageable and the hospital prefers to throw them into municipal bins. General **waste**, which is rich in organic material, is a very strong media for the microorganisms in infectious waste component of hospital **waste.** Thus one would give the microorganisms good environs to multiply if the **wastes** are mixed. Exposure to health care **waste** can result in disease or injury.

Rag pickers who rummage through this **waste** in search of disposables are also at great risk. They get exposed to the **waste** through any open wounds; inhalation of air borne pathogens and pricks by sharps and all this makes them prone to infections. Because of this improper handling, treatment and disposal of the **bio-medical waste** leads to spreading of diseases such as HIV and Hepatitis and infections like ocular, genital and skin infections. There may be anthrax, meningitis, haemorrhagic fevers, septicemia, bacterial and fungal infections.

Components of Bio-medical Waste

The following is a list of materials that are generally considered bio-medical waste

Solids

- *Catheters* and tubes
- Disposable *gowns, masks,* and *scrubs*
- Disposable tools, such as some *scalpels* and surgical staplers

- Medical gloves
- Surgical sutures and staples
- Wound dressings

Liquids

- Blood
- Body fluids and tissues
- Cell, organ and tissue cultures

Sharps

- Blades, such as *razors* or *scalpel* blades
- Lancets
- Materials made of glass, such as *cuvettes* and *slides*
- Metal stylets
- Needles
- Plastic pipettes and tips
- Syringes

Laboratory waste

- Animal carcasses
- Hazardous chemicals with biological components
- Media
- Medicinal plants
- Radioactive material with biological components
- Supernatants

Exceptions

Cadavers, urine, faeces, and cytotoxic drugs are not considered bio-medical waste.

Management of Bio-Medical Waste

Every occupier generating the bio medical waste need to install an appropriate facility in the premises or set up a common facility to ensure requisite treatment of waste by June 30, 2000 in accordance with Schedule-I and in compliance with standards prescribed with Schedule V.

The bio-medical waste needs to be segregated into container/ bags at the point of generation in accordance with Schedule II prior to its storage, transportation, treatment and disposal. The container shall be labelled according to Schedule III. The transporter, operator of a facility shall label the Bio-Medical strictly in accordance with the procedure given in Schedule IV.

Sorting of medical wastes in hospital: At the site where it is generated, biomedical waste is placed in specially labelled bags and containers for removal by bio-medical waste transporters. Other forms of waste should not be mixed with biomedical waste as different rules apply to the treatment of different types of waste. Bio-medical waste is treated by any or a combination of the following methods: incineration; discharge through a sewer or septic system; and steam, chemical, or microwave sterilization. Any tools or equipment that come into contact with potentially infectious material and are not disposable or designed for single use are sterilized in an autoclave.

Household bio-medical waste usually consists of needles and syringes from drugs administered at home (such as insulin), soiled wound dressings, disposable gloves, and bed sheets or other clothes that have come into contact with bodily fluids. Disposing of these materials with regular household garbage puts waste collectors at risk for injury and infection, especially from sharps as they can easily puncture a standard household garbage bag. Many communities have programmes in place for the disposal of household biomedical waste. Some waste treatment facilities also have mail in disposal programmes.

Bio-medical waste treatment facilities are licensed by the local governing body which maintains laws regarding the operation of these facilities. The laws ensure that the general public is protected from contamination of air, soil, groundwater, or municipal water supply.

Training of hospital staff on bio medical waste management is one of our focus areas. As we attempted to resolve particular problems and respond to the queries of the hospital staff, we enhanced our understanding of the practical problems and the unique needs of health care institutions. Apart from training hospital staff, we have also conducted various Training of Trainers (ToT) programmes all around the country, in association with various hospitals and Pollution Control Boards/ Committees.

By the end of such sessions, trainees are exposed to a lot of information, but they do not have enough time to assimilate everything. They have expressed the need for a comprehensive resource on training. This manual has been compiled to fulfil their requirement. The manual has been produced to provide a convenient, up to date training resource that will allow interested people and trainers to increase awareness on waste management and related issues at every level in their organization.

The training manual has six sections and each section has slides on a particular topic. Most of the points in the slides are self-explanatory, but some of them, which may need explanations, have descriptive notes.

Schedule I

Categories of Bio-medical Waste

Waste category	Treatment and disposal
Human anatomical waste	Incineration[a], Deep burial[b]
Animal waste: animal tissues, organs, body parts, carcasses, bleeding parts, fluid, blood and experimental animals used in research, waste generated by veterinary hospitals, colleges, discharge from hospitals, animal houses.	Incineration[a] Deep burial[b]
Microbiology and biotechnology waste: waste from laboratory cultures, stocks or specimens of micro-organisms live or attenuated vaccines, human and animal cell culture used in research and infectious agents from research and industrial laboratories, waste from production of biologicals, toxins, dishes and devices used for transfer of cultures	Local autoclaving/ micro-waving/ incineration [a]
Waste sharps, needles, syringes, scalpels, blades, glass, etc. that may cause punctures and cuts. This includes both used and unused sharps.	Disinfection (chemical treatment): Auto-claving/micro-waving and mutilation/ shredding[d]

Discarded medicines and cytotoxic drugs waste comprising outdated, contaminated and discarded medicines.	Incineration/destruction and drugs[a]. Disposal in secured landfills
Solid waste items contaminated with blood and body fluids including cotton, dressings, soiled plaster casts, lines, beddings, other material contaminated with blood.	Incineration[a]/autoclaving/ micro-waving
Solid waste: waste generated from disposal items other then waste sharps such as tubings, catheters, intravenous sets, etc.	Disinfection by chemical treltment[c] autoclaving/ micro-waving and mutilation/shredding[d]
Liquid waste: waste generated from laboratory and washing, cleaning, housekeeping and disinfecting, activities.	Disinfection by chemical treatment[c] and discharge into drains
Incineration ash from incineration of any bio-medical waste.	Disposal in municipal landfill
Chemical waste: chemicals used in production of biological, chemicals used in disinfection, as insecticides, etc.	Chemical treatment and discharge into drains for liquids and secured landfill for solids

Note: [a]There will be no chemical pre-treatment before incineration. Chlorinated plastics shall not be incinerated.

[b] Deep burial shall be an option available only in towns with population less than half a million and in rural areas.

[c] Chemical treatment using at least one per cent hypochlorite solution or any other equivalent chemical reagent. It must be ensured that chemical treatment ensures disinfection.

[d] Mutilation/shredding must be such that it prevents unauthorized reuse.

Source: 1. Manual on Hospital Wastage Management, 2000, Central Pollution control Board, Ministry of Environment and Forests.

2. *TERI Energy Data Directory & Yearbook,* 2000-01, Tata Energy Research Institute, New Delhi.

Schedule II

Segregation of Bio-medical Waste

1) The segregation of bio-medic waste is the key to successful bio-medical waste management.

2) BMW should not be mixed with any other kind of waste.
3) It should be separated/segregated at the point of generation before storage or transport. The container should be labelled according to the schedule.

Universal Precautions

Universal precautions refers to the practice, in medicine, of avoiding contact with patients' bodily fluids, by means of the wearing of nonporous articles such as medical gloves, goggles, and face shields.

Under universal precautions all patients are considered to be possible carriers of blood-borne pathogens. The guideline recommends wearing gloves when collecting or Handling blood and body fluids contaminated with blood and wearing face shields when there is danger of blood splashing on mucous membranes and when disposing of all needles and sharp objects in puncture resistant containers. Universal precautions are recommended for doctors, nurses, patients, and health care support workers who are required to come into contact with patients or bodily fluids. This includes staff and others *who* may not come into direct contact with patients.

Universal precautions should not be confused with Standard Precautions *which* goes beyond universal precautions. Pathogens fall into two broad categories, bloodborne (carried in the body fluids) and airborne, Universal precautions should be practice in any environment where workers are exposed to bodily fluids, such as:

- Blood
- Semen
- Vaginal secretions
- Synovial fluid
- Amniotic fluid
- Cerebrospinal fluid
- Pleural fluid
- Peritoneal fluid
- Pericardial fluid

Bodily fluids that do not require such precautions include:

- Faeces

- Nasal secretions
- Urine
- Vomitus
- Perspiration
- Sputum
- Saliva

Universal precautions are the infection control techniques that were recommended following the AIDS outbreak in the 1980s. Every patient is treated as if infected and therefore precautions are taken to *minimize* risk. Essentially, universal precautions are good hygiene habits, such as hand washing and the use of gloves and other barriers, correct sharps handling, and aseptic techniques.

Additional precautions are used in addition to universal precautions for patients who are known or suspected to have an infectious condition, and vary depending on the infection control needs of that patient. Additional precautions are not needed for blood borne infections, unless there are complicating factors.

Conditions indicating additional precautions:

- Prion diseases (e.g. Creutzfeldt Jakob disease)
- Diseases with air borne transmission (e.g. tuberculosis)
- Diseases with droplet transmission (e.g. mumps, rubella, influenza, pertussis)
- Transmission by direct or indirect contact with dried skin (e.g. colonization with MRSA) or contaminated surfaces

or any combination of the above.

Protective clothing includes but is not limited to:

- Barrier gowns
- Gloves
- Eyewear (goggles or glasses)
- Face shields

Bio-Medical Waste (Management and Handling) Rules, 1998

With a view to control the indiscriminate disposal of hospital waste/ bio-medical waste, the Ministry of Environment & Forest, Government of India has issued a notification on Bio-Medical Waste Management under the Environment (Protection) Act. Government of NCT of Delhi in its notification dated July 6, 1999 has authorized Delhi Pollution Control Committee (DPCC) for the purpose of granting authorization for collection, reception, storage, treatment and disposal of bio-medical waste to implement the Bio-Medical, Waste Management Rules, 1998. Government of NCT of Delhi has also constituted advisory committee, appellate authority in exercise of powers conferred under Bio-Medical rules.

Some of the salient features of these rules are:

Rules are Applicable to: These Rules will apply to hospitals, Nursing Homes, Veterinary Hospitals, Animal Houses, Pathological labs and blood banks, generating hospital wastes. *(except such occupier of clinics, dispensaries, pathological labs, blood banks providing treatment/ service to less than 1000 (one thousand) patients per month).*

Duty: It shall be the duty of the every occupier of an institution generating bio medical waste which includes a hospital, nursing home, clinic, dispensary, Veterinary institution animal house, pathological laboratory, blood bank by whatever name called to take all steps to ensure that such waste is handled without any adverse effect to the human health and the environment.

Penalty: The defaulting hospitals/nursing homes, etc. are liable to be penalized as per the provisions of Environment (Protection) Act, 1986 and other Pollution Control Acts.

Appeal: Any person aggrieved by an order made by the DPCC under these rules may within thirty days from the date on which the order is communicated to him, prefer an appeal to the Financial Commissioner, Government of NCT of Delhi who is appointed as Appellate Authority under the rules.

Management Based on the Categories of Bio-Medical Waste

Chemicals treatment using at least 1% hypochlorite solution or any other equivalent chemical reagent and it must be ensured that chemical treatment ensures disinfections.

+Mutilation/shredding must be such so as to prevent unauthorized reuse.

*There will be no chemical pretreatment before incineration. Chlorinated plastics shall not be incinerated.

•Deep burial shall be an option available only in towns with population less than five lakhs and in rural areas.

Treatment and Disposal of the Waste Category

Option	*Treatment and Disposal*	*Waste Category*
Cat. No.1	Incineration *deep• burial organs, body arts)	Human Anatomical Waste (human tissues, organs, body parts)
Cat. No.2	Incineration */deep burial	Animal waste animal tissues, organs, body parts carcasses, bleeding parts, fluid, blood and experimental animals used in research, waste generated by veterinary hospitals/colleges, discharge from hospitals, animal houses)
Cat.No.3	Local autoclaving/ micro waving /incineration*	Microbiology and Biotechnology waste (wastes from laboratory cultures, stocks or specimens of micro organisms live or attenuated vaccines, human and animal cell culture used in research and infectious agents from research and industrial laboratories, wastes from production of biological, toxins, dishes and devices used for transfer of cultures)
Cat.No.4	Disinfections (chemical treatment/ autoclaving/micro waving and mutilation sharps) shredding	Waste sharps (needles, syringes, scalpels blades, glass, etc. that may cause puncture and cuts. This includes both used and unused
Cat.No.5	Incineration */ destruction and disposal in secured landfills	Discarded medicines and cytotoxic drugs (wastes comprising outdated, contaminated and discarded medicines)
Cat. No.6	Incineration *, autoclaving/ micro waving	Solid waste (Items contaminated with blood and body fluids includingcotton, dressings, soiled plaster casts, line bedding beddings, other material contaminated with blood)

Cat.No.7	Disinfections by chemical treatment treatment autoclaving/micro waving & mutilation shredding	Solid waste (waste generated from disposable items other than the waste sharps such as tubing, catheters, intravenous sets, etc.)
Cat.No.8	Disinfections by chemical treatment and discharge into drain	Liquid waste (waste generated from laboratory and washing, cleaning, housekeeping in and disinfection activities)
Cat.No.9	Disposal in municipal landfill	Incineration ash (ash from incineration of any bio-medical waste)
Cat.No.10	Chemical treatment and discharge into drain for liquid and secured landfill for solids	Chemical waste (chemicals used in production of biological, chemicals, used in disinfection, as insecticides, etc)

The most essential part of hospital waste management is the segregation of Bio medical waste. The segregation of the waste should be performed within the premises of the hospital/nursing homes.

Color Coding and Type of Container for Disposal of Bio-Medical Waste

The colour coding, type of container to be used for different waste categories and suggested treatment options are listed below.

Colour Coding	*Type of Containers*	*Waste Category*	*Treatment Options as per Schedule 1*
Yellow	Plastic bag	1,2,3,6	Incineration/deep burial
Red	Disinfected Container/ Plastic bag	3,6,7	Autoclaving/ Micro waving/ Chemical Treatment
Blue/ White translucent	Plastic bag/puncture proof container	4,7	Autoclaving/Micro waving/chemical treatment and destruction/shredding
Black	Plastic bag	5,9,10 (Solid)	Disposal in secured landfill

Notes:

1. Colour coding of waste categories with multiple treatment options as defined in ***Schedule 1,*** shall be selected depending on treatment option chosen, which shall be as specified in Schedule 1.
2. Waste collection bags for waste types needing incineration shall not be made of chlorinated plastics.
3. Categories 8 and 10 (liquid) do not require containers/bags.
4. Category 3 if disinfected locally need not be put in containers/bags.

Different labels for bio-medical waste containers and bags shall be required for identification and safe handling of this waste. These labels for storage/ transportation of bio-medical waste are as under,:

Schedule III

Label for Bio-Medical Waste Containers/Bags

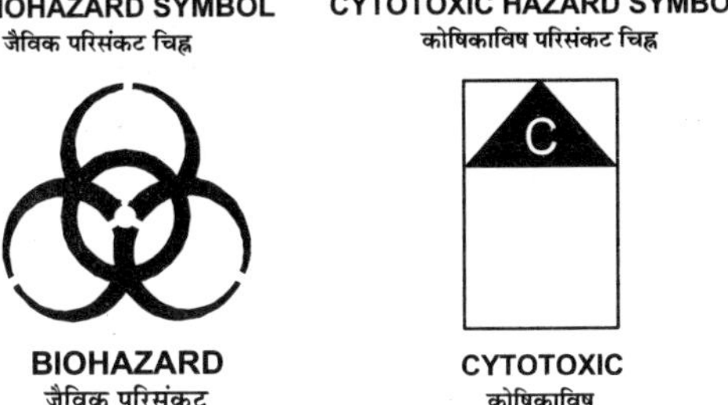

Schedule IV

Transportation

1) Each health care facility should have a health care waste management plan which should include collection points and routes of waste transport
2) A time table of the frequency of collection should also be set up.

Bags to be filled with only 2/3rd capacity, bags should be sealed/ labelled from Source of Generation

Date of Production, Place of Production Waste Category with Signature should be mentioned

* Untreated BMW should only be transported in a vehicle as may be authorized for the purpose by the competent authority.

* No untreated BMW should be kept or stored beyond a period of 48 hrs.
* If it needs to be stored beyond this period then an authorized person should take permission from the competent authority.
* Biomedical waste should be transported within the hospital by means of wheeled trolleys that are not used for any other purpose.
* These trolleys should not have sharp edges and should be cleaned daily. *Biohazard symbol should be painted on the trolley.
* No Personal Protective Clothing, no Overloading of Trolley, preventing excessive Spillage of Waste and Transportation of Bio-medical waste is away from Patient care units
* Notwithstanding anything contained in the Motors Vehicles Act, 1998 or rinds there under. Bio-medical waste shall be transported only in such vehicles as may be authorized for the purpose by the competent authority

Schedule V

Standards prescribed

PPE: Personal Protective Equipment

* Provide health care staff with gloves, boots, aprons, masks, goggles, hand gloves, head caps, face masks, safety shoes, latex surgical gloves, heavy duty rubber gloves, nitrile and neoprene gloves, dust masks, plastic aprons, protective glasses, gumboots, immunization/Hepatitis B/tetanus/ typhoid for personal protection
* Educate and encourage the health care staff to use these PPE and report injuries as and when they occur.

Conclusions

The scenario of bio-medical waste demands the better management. The present rules are ineffective to manage the bio-medical waste. The law relating to bio-medical waste management should play an important part in curbing the menace. It is sad to note that the surveys itself reveals poor implementation of the bio-medical rules...

The law has stumbled in performing its duty; much of attribute from the lack of proper implementation mechanism Though the law relating to bio-medical waste is in its infant stage, the time has come to act seriously and implement the rules effectively. Greater commitment is required on the part of the government looking into the magnitude of the problem. The regulatory body should be strengthened. There is a need for a reappraisal of the rules.

The rules must drop the plan of imposition of setting up of a treatment plant in every hospital. The rules should devote its attention towards establishment of common treatment sites* which includes incinerator or autoclave, shredder and an engineered pit despite the fact that the occupier/operator has the means and potential to handle the same. The rules must be strictly enforced with proper segregation is the secret of proper regulation bio-medical waste. Training of these personnel will prepare the road in that direction. Since the hospitals have become the source of huge profits, they owe a duty to protect the interest and the safety of its workers and the public. Hospitals may face an excessive burden because of regulation of medical waste. But they should realize that the law will no longer tolerate any infringements.

11

Status of Soil Fertility in Core Area of Kanha Tiger Reserve, Madhya Pradesh

Ramendra Tiwari, A.K. Mudgal and K.S. Sengar

Introduction

Soil is the important component of earth crust, which supports the life processes. Soil develops from weathered rocks, volcano or decomposition of organic residues. It is a natural medium for the growth and in case of terrestrial ecosystem it plays the crucial role for the development of flora and micro fauna also. This truth that soil is divided into different sub regions but normally top soil is important for physiological activity for development of plants. Plants get macro and micronutrients from soil as well water. The Kanha tiger Reserve is an excellent interspersion of the Dadars (flat hill tops), grassy expanses, dense forests and riverine forests. It is very rich in flora, largely due to the combination of landforms and soil types, apart from the moist character of the region

As one of the Tiger Reserves in India, Kanha Tiger Reserve is of great potential and importance of Central Indian Highlands. Kanha Tiger Reserve falls in Balaghat and Mandla districts of M.P., dominated by Sal with multitude of deer, antelopes, Gaurs and tigers. The protected area is divided into two zones, Core and Buffer.

The entire National Park area of 940 sq. km. is visualized as a "core zone", free from all biotic disturbances. A buffer zone of 1009 sq. km. surrounding the core is treated as a "multiple use area". The buffer zone comprises almost 40 per cent of forest area and the rest is constituted by revenue and private land. This zone is characterized by an interspersion of revenue and forestlands, pockmarked with numerous villages. Besides, the Phen Wildlife Sanctuary, a satellitic micro core of 110 sq. kms, is also under the administration of the Reserve Management. The significance of this micro core lies in the exchange of gene pool and dispersal of wildlife populations between the two conservation units. The buffer zone has been notified by the state Government as a separate division and is under the unified control of the Reserve Management. Wildlife conservation practices and eco development in the National Park, Buffer Zone and Phen Wildlife Sanctuary are carried out under the Management Plan for the year 2001-2002 to 2010-2011. The major fauna is Chital, Sambar, Barasingha, Barking deer, Chousingha, Gaur, Langur, Wild pig, Jackal, Sloth bear, Wild dog, Panther, Tiger. To understand the role of lithosphere in development and sustainability of an ecosystem as important as a tiger reserve, present study of soil will be useful. This paper deals with the core zone of 940 km2 area of National Park. The core area shows black cotton, alluvium, sahara and barra.

Materials and Method

To carry out the physio chemical analysis of soil, 7 sample sites were selected and samples were taken using standard methods. Samples were brought to the laboratory in sterile conditions and were processed for the physico chemical analysis. Soil samples were analysed for pH, conductivity, N, P and K. For pH standard method using pH meter with 1:2 ratio dilution was used. Electrical conductance was measured with the same sample with conductivity meter. Organic carbon was determined by the Walkey and Black method. Nitrogen was analyzed using micro Kjeldhal method where as Phosphorus was done by the Bray and Olsen method. Potassium was analysed using the flame photometer method.

The sites selected were:

	SITE	AREA
1.	I	Kanha meadow (Kanha Range)
2.	II	Kanha forest (Kanha Range)
3.	III	Bamhni Dadar meadow (Kanha Range)
4.	IV	Sondhar meadow (Mukki Range)
5.	V	Bisanpura forest (Mukki Range)
6.	VI	Kisli meadow (Kisli Range)
7.	VII	Kisli forest (Kisli Range)

Results and Discussion

The present study reveals the nutrient status of the core zone of Kanha Tiger Reserve, which is under natural conditions without any input of artificial fertilizers. pH in SITE I, II, VII show neutral pH, while SITE III, IV ,V are slightly acidic, while SITE VI is acidic. According to Raina et al (2000) average pH in the surface, soil is 7.45 the low pH shows high organic matter content. Similar results have been shown by Horkar et al. (2002) Navegaon National Park, Maharashtra.

Electrical conductance shows no considerable differences in all the cases it is less than 1 considered to be normal. The SITE III and IV show least electrical conductance.

Organic carbon observation is very low from SITE II and V, while SITE I shows low organic content where as SITE III, IV and VII show high concentration.

The bulk of the nitrogenous material found in soil is added from plant residues is organic and largely unavailable. In the present study SITE I, II, V show low concentration, SITE III, IV, VI, VII have considerably high nitrogen concentration. The nitrogenous compounds present in soil organic fraction persist to attack being 50% appreciable that only a small proportion of nitrogen reservoir of soil is mineralized in each growing season.

SITE I, III, IV show low value of phosphorus while the rest of the site medium value of phosphorus. The chief source of organic phosphorus is vegetation that undergoes decay. Singh et al. (2001) shows that available phosphorus increase after burning but decline during cropping.

Potassium contents of SITE IV, V, VI, VII is low while the rest is medium clearly indicates that not much microbial transformation of this element has taken place.

The present study indicates a great diversity in meadow and forest regions, which is due to diversity of vegetation and mineralization processes. Meadow being rich in organic contents and good conductivity while phosphorus and nitrogen contents are better in forest areas because of better decomposition processes.

Table 1 : Nutrient Status

SITE	*pH*	*Conductivity (milliohms)*	*Organic Carbon (%)*	*N (%)*	*P (kg/hac)*	*K (kghac)*
I	7	0.12	0.5	0.03	9	371
II	7	0.05	0.3	0.03	10	281
III	6.9	0.04	1.2	0.07	8	213
IV	6.9	0.04	1.4	0.09	8	146
V	6.9	0.14	0.2	0.02	12	146
VI	6.5	0.15	0.9	0.06	10	101
VII	7	0.15	1.2	0.07	12	168

REFERENCES

1. Dimri, B.M., Jha, M.N., Gupta, M.K. (1997). Status of Soil Nitrogen at Different Altitudes in Garhwal Himalaya. Van Vigyan. 35(2): 77-84
2. Duvigneaud, P. and Denaeyer DeSmet, S. (1973) Biological Cycling of Minerals in Temperate Deciduous Forest in: Recihle, D.E. (ed) Analysis of Temperate Forest Ecosystem. 199-225. Springer Verlag, Berlin.
3. Garcia, Oliva F., Sanford, R.L. Jr., Kelly, E. (1999). Effect of Slash and Burn Management on Soil Aggregate Organic C and N in a Tropical Deciduous Forest. Geoderma 88: 1-2, 1-12, 32.
4 George E., Stober. C., Seith, B. (1999). The Use of Different Soil Nitrogen Sources by Young Norway Spruce Plants. Trees: Structure and Function 13:4, 199-205.
5 Hanson, H.C. (1950). Ecology of Grasslands II. Bot. Rev. 16:6.
6 Horkar, V.M., and N.G. Totey (2002). Characterization of Soils of

Navegaon National Park (Maharashtra). *Indian Journal of Forestry.* Vol. 25 (2) :127-135.

7 Incerti, M., Clinnick, P. F., Willatt, S.T. (1987). Change in the Physical Properties of a Forest Soil Following Logging. Australian Forest-Research. 17(2): 91-98.

8. Jahn, R.; Asio, V.b.; Schulte, A. and Ruhyiyat, D. (1998) Soil of the Tropical Forest of Leyte Philippines I: Weathering Soil Characteristics, Classification and Side Qualities. International Conference, Balikpapan, Indonesia. 29-36 pp.

9. James, B.R., Aschmann, S.G. (1992). Soluble Phosphorus in a Forest Soil Ap Horizon Amended with Municipal Westwater Sludge or Compost. *Communications in Soil Science and Plant Analysis.* 23 : 7-8, 861-865.

10. Koba K., Tokuchi N., Yoshioka T., Hobbie E.A., Iwatsubo G. (1998) Natural Abundance of Nitrogen 15 in a Forest Soil. *Soil Science Society of America* Journal 62:3, 778-781.

11. Mishra, D.; T.K. Mishra and S.K.Banerjee (1997). Comparative Phyto-sociological and Soil Physio Chemical Aspects between Managed and Unmanaged Lateritic Land. *Ann. For.* 5(1) 16 25.

12. Nye, P.H. (1961). Organic Matter and Nutrient Cycles under Moist Tropical Forest. *Plant and Soil,* 13:333-346.

13. Puri, A.N. and Asgher, H.G. (1940) Physical Characteristic of Soil. Vll. Effect of Ignition; *Soil science* 49: 369-378.

14. Purwanto, Gintings, A.N. (1994). Research on the Physical and Chemical Soil Properties under Forest Stands of Dua Banga Moluccana in West Nusa Tenggara. *Bulletin Penelitan Hutan.* 561: 1-24.

15. Purwanto, Gintings, A.N. (1994). Research on the Physical and Chemical Soil Properties under Forest Stands of Dua banga Moluccana in West Nusa Tenggara. *Bulletin Penelitan Hutan.* 561:1 24. Quality. Ph.D. Thesis. Univ of Saugor Sagar.

16. Raina, J.N. and Pradeep Kumar (2000). Physiochemical Characterization and Fertility Status of Some Forest Nursery Soil of District Sirmour in Himachal Pradesh. *Indian Forest.* 126:657-663

17. Singh, J., Indrani P. Bora and A. Baruah (2001). Physio Chemical Attributes of Soil under Jhum Cultivation in Ampehngiri (Burnihat), *Meghalaya. Ann. For.* 9(2):257-263.

12

Environmental-Friendly Biocidal Activity of Copper Surfactants Derived from Non-edible Oils (i.e. Neem and Karanj) and Their Benzothiazole Complexes

Shema Khan, A.R. Khan and Rashmi Sharma

Copper soaps derived from Neem *(Azadirecta indica)* and Karanj (Pon *gamia pinnata)* oils and their benzothiazole complexes have been synthesized using direct metathesis. These newly synthesized compounds were characterized on the basis of elemental and spectral analysis evaluated for biological properties. In the present work we report the fungicidal activity of these copper soaps and their complexes against fungi.

Introduction

Copper soaps derived from non edible oils i.e. Neem and Karanj play a vital role in various fields due to their surface active properties[1,2]. These oils were particularly chosen as they themselves possess anti-fungal activity, can easily be extracted and biodegradable in nature. Copper soaps have a tendency of complexation with 'Nitrogen' and 'Sulphur' containing ligands. Using benzothiazole ligand for complexation of copper soaps have been done to obtain their benzothiazole complexes. Since copper metal is toxic in nature, literature survey reveals that the synthesized copper soaps and their benzothiazole complexes may play a significant role in biological activities and have sufficient pharmaceutical, industrial and analytical applications[3,4,5]. Biocidal studies have been done to assess their comparative toxicity on easily available fungi *Alternaria alternate.* The studies lead us to the conclusion that copper soap and their complexes may play a significant role in various fields of industries, wood preservation and agrochemical etc.

Experimental

Copper soaps derived from Neem and Karanj oil and their benzothiazole complexes have been synthesized and studied for structural aspects. All chemicals used were of LR/ AR grade. Neem and Karanj oils were extracted from their kernels respectively using petroleum ether and then purified. The fatty acid composition of these non edible Oils was confirmed through GLC of their methyl esters and is given in Table 1. Copper soap was prepared by refluxing the non edible oils i.e. Neem and Karanj oil with ethyl alcohol and 2 N KOH solutions for 3-4 hours (Direct Metathesis). The neutralization of excess of KOH present was done by slow addition of 0.5N HCI. Saturated solution of Copper Sulphate was then added to it, for conversion of neutralized potassium soap into their corresponding copper soaps. Copper soap so obtained was then washed with warm water and 10% alcohol at 50°C and recrystallized using hot benzene.

Metal was analysed by standard procedure (iodometrically).[6] The purified copper soap derived from non edible oil, i.e. Neem and Karanj were refluxed separately with ligand 2 amino 6 methyl benzothiazole in 1:1 ratio. Ligand was dissolved in 2 3ml of hot ethyl alcohol and copper soaps were dissolved in 10-15 ml of Benzene separately and both solutions were added with stirring. The above reacting mixture was them refluxed for 3-4 hours with using water condenser. Further solid product so formed, was filtered hot, dried and recrystallized and purified. TLC using silica gel was used to check the purity of the complex. Their molecular weights were determined as similarly as calculated for average molecular weight of soaps.

The complexes are abbreviated as follows:

Copper—Neem soap (CN).

Copper—Pongamia soap (CP).

Copper—Neem benzothiazole complex (CNB).

Copper—Pongarnia benzothiazole complex (CPB).

All the soap and complexes are stable at room temperature, their physical parameter are reported in Table II. On the basis of their elemental analysis, 1:1 (metal: ligand) type of stoichiometry has been suggested. In order to study the structure of soaps, the Infra red absorption spectra of compounds were obtained on a FTIR spectrophotometer, Shimadzu 821PC (4000-400 cm^{-1}) from sophisticated Analytical Instrument Facility, CDRI, Lucknow. Proton NMR spectra were also recorded at SAIF, CDRI, Lucknow on NMR spectrometers, Bruker DRX 300 at 300K using C_6 D_6 (deuterated benzene) as solvent for soaps. They were examined to determine bonding and tautomerism present in the structure.

The antifungal activities of oils, copper soaps and their urea complexes have been evaluated by testing against easily available fungi *Alternaria alternata* which was isolated from natural habitat and then purified, characterized and identified. The antifungal activity of Copper Soaps and their complexes were checked by the agar plate technique. The culture medium used for the growth of the organism was natural media, i.e. P.D.A. suggested by Booth and Hawksworth?[7,8]. The photograph of natural habitat and culture of fungi Alternaria alternate is shown in Fig. IA and IB respectively. The solution was prepared in ppm of different concentration (i.e. 10^3 and 10^4 PPM) in methanol benzene mixture. The effect of solvent has been excluded by complete evaporation. After the solidification the above medium and evaporation of solvent, single fungal hypha/spore were isolated and transferred on agar slants (1%) for multiplication. *Serial dilution method* was used.

Fungicidal Testing: Firstly 2ml. of the fungal broth was poured in sterilized Petri-plates and than the sterilized P.D.A. media was poured and after the gel was set it was dried with the lids agar till moisture has evaporated, i.e. the surface of agar was kept moist but with no droplets. Into these plates 5 ml of the diluted testing solution was aseptically transferred by rotating the Petri plates in clockwise and anti clock wise direction 3-4 times and was allowed incubate at 25±1° C for 24 and 48 hours. After the period of incubation the plates were observed for the growth of fungus in different concentration of the testing solution used in the present study. The data were statistically analysed according to the following formula.

% inhibition = (C T)/C * 100

C = number of fungal colonies in control plate.

T = number of fungal colonies, in test plate.

Results and Discussion

In the IR spectra of copper soaps a band in the region 750-450 cm^{-1} are due to copper oxygen (Cu-O) bond, which are characteristic of absorption of metal constitutent of each soap molecules, whereas the strong band appears in the region 2970-2840 cm^{-1} confirms the methyl groups in acids of various chain length. In the NMR spectra of copper soap benzothiazole complexes, a peak observed at δ-3-2–4.2 confirms the NH_2 group of benzothiazole segment to the metal atom of the soap segment. The antifungal activities of copper soaps and their complexes for *Alternaria alternata* was checked by agar plate technique.

Each plate was examined after 24 and 48 hours of incubation. The number of fungal colonies were counted using colony counter. The % inhibition was calculated according to the given formula above.

It is evident from the results that concentration plays a vital role in increasing the degree of inhibition. Fungicidal screening data revealed that at lower concentration the inhibition of growth is less as compared to the higher concentration of the complexes. And so the field is still open for further research and analysis can be made for a longer period of incubation and the work can be extended. Chaurasia et a1.[9] studied anti-fungal activity of transition metal complexes of 6-methyl-2 amino benzothiazole. The CuL_2I_2, Cu L_4I_2, HgL_2I_2 and Hg L_4I_2 (where L = 2-amino-6-methyl benzothiazole) were screened for antifungal activity on Aspergillus niger, Alternaria alternate, Curvularia plunata and Penicillium fumculorus by employing Potato dextrose agar method at two dilutions (1:2143 & 1:5000).The results indicate that the activity of Cu (II) and Hg (II) complexes against *Alternaria alternata* is of a very high order (100%), whereas that of *Aspergillus niger* is only 91 per cent. Cu $L_4 1_2$ Inhibits 100 percent and 85 percent of spore germination of *Pencillium fumculorus, Culvularia plunata* respectively. Earlier workers suggested that complexes are more effectively playing their role to check the growth of fungi[10, 11]. Complexes are better antifungal agents than soaps and the Neem complex is more effective than the Pongamia complex for 48 hours (Figure 11). The N donor ligand used during the present investigation is 2-amino 6 methyl benzothiazole, which has been very well reported to possess biocidal activity. All the above results indicate that 'Nitrogen' and 'Oxygen' containing ligands enhance the fungicidal activity of the soap molecule. These studies will play a significant role in selection and promotion of eco-friendly and biodegradable fungicides, pesticides and insecticides. The enhanced activity of synthesized complexes as compared to those of the soap can be seen and possibly be explained on the basis of chelate formation, presence of donor atoms, basicity as well as the structural compatibility with molecular nature of the toxic moiety. Enhanced biological activity of complexes is in accordance with the chelation theory. [12]

Table 1: Percentage Fatty Acid Composition of Oils Used for Preparation of Copper Soap/Complex.

	Percentage Fatty Acid (Carbon Number)							
Name of Oil	16:0	18.0	18.1	18.2	20.0	20.1	22.0	24.0
NEEM Oil	14.9	14.4	61.9	7.5	1.3	-	-	-
KARANJ Oil	5.2	5.0	57.3	13.8	2.7	10.3	4.1	1.6

Table II : Analytical and Physical Data of Copper Soaps and Complexes Derived from Neem/Pongamia

Name of Soap/ Complex	*Colour*	*M.Pt. (°C)*	*Metal Content Observed*	*Metal Content Calculated*	*S.V.*	*S.E.*	*Average Mol. Wt.*
CN	Dark green	50	10.3632	10.1087	198	283.33	628.166
CP	Dark green	51	9.4996	9.3426	181.5	309.091	679.682
CNB	Greenish brown	63	7.7112	7.9983	-	-	793.918
CPB	Greenish brown	65	7.3406	7.5109	-	-	845.434

Figure II: Anti-fungal Activity of Neem and Pongamia Oils, Their Copper(II) Soaps, Their Complexes and Benazothiazole Ligand *after 48 Hours* of Incubation for *Alternaria alternata*

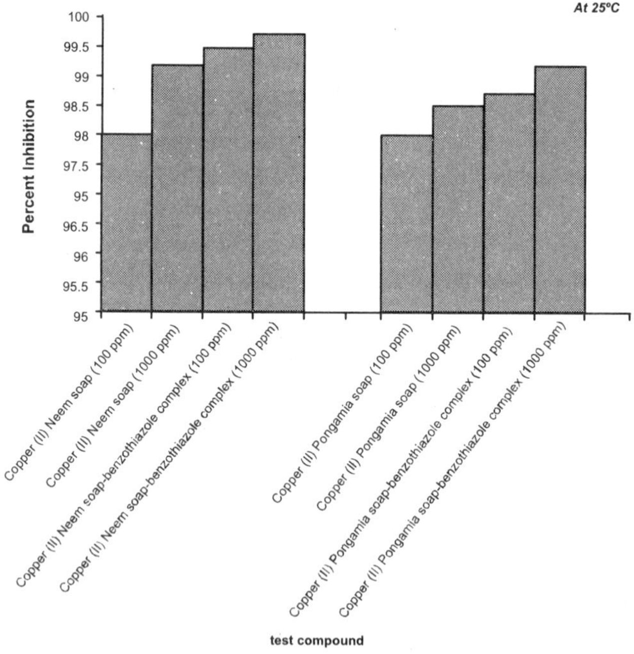

Oils show 100% inhibition of fungus.

REFERENCES

1. Sharma R. and Khan, RPMP Natural Products, 18, 429-435, (2007)
2. Sherwani, MRK, Sharma R., Gangwal, A. and 2527, Bhutra R., Indian J. Singh N., Sangwan N.K. and Dhindsu K.S., Pest Manag, Sci., 56, 28, (2000). Mukherjee G.N. and Das A., J. *Indian Chem, Soc.,* 78, 78 (2001).
3. Gunstone, F.D., An Introduction of the Chemistry of Fats and Fatty Acids, Chapman & Hall Ltd. London, 1958.
4. Horsfall. J.G., Bot. Rev. 11, 357, (1945).
5. Hingo, Alsushi and Veda Hiroshi Nishil, Kokai Tokkyo Koho IP. 09, 249, 654 (1997).
6. Prince, F.M. and Sansford, K.E., Tissue Culture Asso. Manual, 2. 379. (1976).
7. Booth, C. Methods in Microbiology, Acad Press. N.Y., 4, (1971), 795.
8. Hawks worth O.L. Mycologists Hand Book: An introduction to the Principles of Taxonomy and Nomenclature in the Fungi and Lichens, C.M.I. Kew, England (1974).
9. Chaurasia, M.R. Shukla P. and Singh, N.K. Def Sci J, Vol. 32, No. 2, (1982) 75-99
10. Mehta, V.P. , Talesara P.R. and Sharma R.J. *Indian of Chem. Soc.,* 39A, (2001), 383
11 Mehta, V.P., Talesara P.R. and Sharma R.J. *Indian of Chem.* Soc.,40A, (2000), 399.
12. Srivastava, R.S. Inorg. Chim. Acta 151, (1988), 285.

Acknowledgments

The authors would like to thank UGC, for financial support, Principal and Head of the Department of Chemistry, S.D. Government College, Beawar for providing Laboratory Facilities.

13

Impact of Exotic Species Invasion on Indigenous Plant Biodiversity

Sudhir Pathak, S.D. Rathor and S.N. Sharma

Introduction

Diversity is one of the most important characteristics of nature which pervades the whole universe. Change is the law of nature and bio-diversity is the natural beauty and natural resources of all living things. Bio-diversity is a collective term that encompasses the variety of all plants, animals and micro-organisms on earth. Bio-diversity can be divided into three categories genes, species and ecosystems. These types of bio-diversity provides stability of earth. India is one of the 12 mega bio-diversity countries in the world. The ministry of environment and forests, Government of India (2000;1 records 47,000 species of plants and 81,000 species of animals which is about 7% and 6.5% respectively of global flora and fauna. In India loss of bio-diversity causes many serious problems creates like disaster, Acid rain, Tsunami, Global warming, environmental pollution, increasing sea level and extinct of many plant and animal species of the world.

Destruction and loss of natural habitat is the single larger cause of bio-diversity loss. Billions of hectares of forests and grasslands have been cleared over the past 10,000 years for conversion into agriculture lands, pastures, settlement areas or development projects. These natural forests and grasslands were the natural homes of thousands of species which perished due to loss of their natural habitat. Many factors responsible for loss of Bio-diversity but exotic species invasion are the one of the most important factor which affect the Bio-diversity. Invasion of exotic species is among the most important global scale problems. The spread of these species threatens ecosystem, habitats or species with economic and environmental harm. Over 40% of the species on the list of threatened and endangered species are due to invasive species. Although large number of exotic species have become naturalized in India and have affected the distribution of indigenous flora and vegetation patterns of the country.

Lantana camara, Parthenium hysterophorus, Prosopis juliflora, Mimosa invisa, Eupatorium adenophrum, chromolaena odorata and **cytisus scoparius** among terrestrial exotics and *Eichornia crassipes* (water hycianth) and *Pistia stratiotes* among aquatics have posed serious threat to the native indigenous flora.

The main objective of this paper is to focus on how the exotic species' invasion spread, its biological characteristics and effects on indigenous flora, and economy of a region.

Exotic Species Invasion Mechanism

Invasion by an alien plant is its introduction to an area. Introduction of non-indigenous species may occur through (i) accidental introduction and (ii) Importation.

Once introduced, the invader formation of colonies the new habitat, produces self populations into the natural flora.

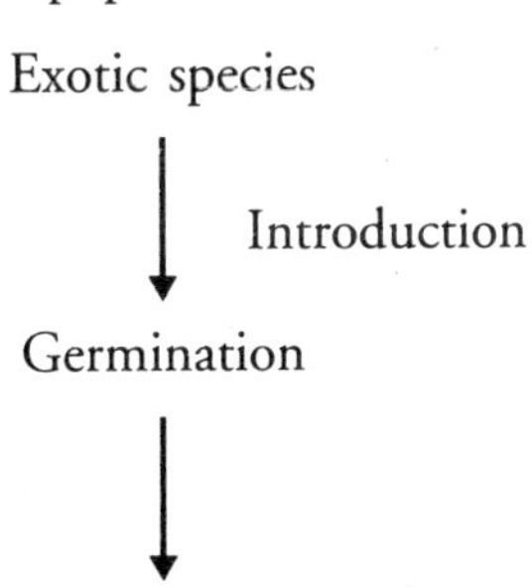

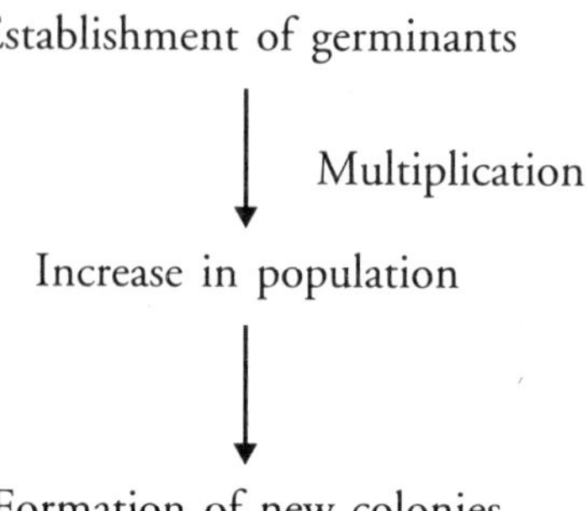

Impact of Exotic Species on Community and Economy

The impact of exotic plant species on community structure, hydrobiological cycle, ecosystem nutrients, food web, food chain and energy flow. *Lantana camara* has spread to Pachmarhi biosphere reserves in M.P. and Nainital in U.P. *Parthenium hysterophorus* has also spread all over India and the currently infested area is estimated at 20,25,000 hac. *Lantana* infests 4 m. hac. in Australia and it has also infested millions *of* hectares in 47 countries excluding India *Eichhornia crassipes* (water hyacinth) was introduced into India in 1896 as an ornamental pond plant. Now this weed is infecting more than 2000,000 hac. of water surface causing concern in 98 out *of* 246 districts in India. It interferes with production of hydroelectricity, blocks water flow in irrigation and drainage canals. It also affects the aquatic fauna through elimination of habitat and depletion of oxygen level caused by respiration and decomposition of vegetative parts.

The economic impact of exotic species are both direct and indirect. In 2001 FAO identified six types of economic impacts of plant invasion. (i) On production (ii) On price and marketing effects. (iii) on Business (iv) on Nutrition (v) on human health and environment

Control Measures

For control of the exotic invasive plant species, manual, mechanical, chemical and biological control methods may be applied. The necessary research work and effective control measures to control invasive plant species are in progress. The governments and people related to the concern to preventive the introduction of control through legislative element and eradication programme those alien species which threaten ecosystem habitat or species. Manual removal is a labour intensive and low efficiency technique. Mechanical control involves generally mechanized equipments. Chemically control involved the use of inorganic and organic herbicides. Biologically control programmes is not clear cut. In India the biocontrol agent (Teleonemia scrupulosa) relased for *Lantana* control failed since the control agent could not cope with the vigorous regrowth of *Lantana* at the onset of monsoon rains.

In the case of *Eichhornia crass1pes* the most widely followed practice among the people is by manual collection and destruction. *Fusarium pallidoroseum,* a wilt inducing pathogen isolated from water hyacinth could restrict the multiplication and thus cause reduction in the population of the weed.

Conclusion

Exotic species invasion is the most critical problem of India. India is a rural country and 75% people live in rural areas and agriculture is the main source of economy, so there is an urgent need of studies on biological invasions in India.

REFERENCES

Aneja, K.R. Dhawan, S.R. and Sharma, A.B., Deadly Weed Parthenium Hysterophorus L. and its Distribution. *Indian J. Weed Sci* 1991, 23, 14-18.

Gyan P. Sharma, J.S. Singh and A.S. Raghubanshi, Plant Invasions: Emerging Trends and Future Implications, *Current Science*, Vol. 88, No. 5, March 10, 2005.

Mantri, A. Annapurna, C and Singh J.S. Terrestrial Plant Invasion and Global Change. In Bioresource and Environment (eds Tripathi, G. and Tripathi, Y.C.), Campus Book International, New Delhi, 2002 pp. 25-44.

Naseema, A.R., Praveena, R. RehNair and C.K. Peethambaran, **Fusarium**

pallidoroseum for Management of Water Hyacianth. *Current Science* Vol. 86, No. 6, March 25, 2004.

Pachmarhi Biosphere Reserve, Project Document Environmental Planning and Co-ordination Organization Paryavaran Parisar, E 5 Arera Colony, Bhopal, 1996.

Pathak, S.K. Plant Diversity and Community Patterns of Tropical Evergreen Forest Pachmarhi Hills (M.P.) Ph.D. Thesis IEMPS Vikram University Ujjain (M.P.) 2001.

Ramakrishnan, P.S. (ed.) Ecology of Biological Invasion in the Tropics, International Scientific Publications. New Delhi, 1991.

14

Sensitivity of *Triticum aestivum* cv.HDM 1553 to Sulphurdioxide and Ameliorating Effect of Urea Spray on SO_2 Exposed Plants

Subhash Chand

Introduction

Sulphur dioxide is produced by burning of fuels, coal combustion and fuel oil combustion. The nature and extent of SO_2 effect on plants depend upon concentration of SO_2, physiological and genetic status of plant and relationship between time and concentration. SO_2 is absorbed in the mesophyll of the leaf through stomata and is phytotoxic to growth, yield, stomatal opening, biochemical and physiological processes of plants. Mild doses of SO_2 causes interveinal chlorotic bleaching of leaves in plants, e.g. alfalfa, cotton and barley. 0.3 to 0.5 ppm for several days on sensitive plants causes chronic injury (Kudesia, 2007). Keeping this in view, the present study has been undertaken to determine the sensitivity of *Triticum aestivum* cv. HDM 1553 to long term exposure to SO_2 and further ameliorating effect of 5% aqueous solution of urea sprayed on SO_2 exposed plants, to reduce the extent of damage caused by SO_2 for wheat plants.

Materials and Methods: The seeds of *Triticum aestivum* cv. HDM 1553 procured from National Seed Corporation, were sown in the experimental plots following standard agronomic practices. 10 days old plants were exposed to 667 and 1334 μgm^{-3} SO_2 for 6 hours each day (10 am – 4 pm) up to 110 days in 1 m^3 polythene chambers supported by aluminium frames. SO_2 was generated by passing a continuous current of air (l litre per minute) through an aqueous Sodium Metabisulphite solution, which was ionized under pressure to produce SO_2 (Agarwal et al., 1982). The gas was introduced within the chamber along with an additional flow of air (50 liter per minute) through the perforated alkathane pipes for uniform distribution of gas within the chamber. The concentration of SO_2 in the fumigation chamber was monitored over intervals of 2 hours, sucking a known volume of SO_2 from above crop canopy and observing it into 0.1 m solution of Sodium tetra chloro mercurate which was calorimetrically assessed for SO_2 concentration (West and Gaeke, 1956). A control was run in identical conditions but without any SO_2 fumigation. In two replicates from each concentration 5% aqueous solution of urea sprayed at an interval of 30 days. Samples were collected at intervals for 20 days.

Observations were made on visible injury, leaf area reduction, growth parameters such as plant height, shoot and root length, number of leaves, roots and tillers, biomass, dry weight fractions and net primary productivity (NPP). Data on yield were recorded at the harvest of crop. Plants were also analysed for leaf extract pH, ascorbic acid, phosphorus, sulphur and carbohydrate contents. The leaf pH was determined with a pH meter by homogenizing 5 gram fresh leaves with 25 ml double distilled water. The following methods of analysis were used for their respective analysis: ascorbic acid (Killer and Schwagar, 1977), phosphorus (Jackson, 1958), sulphur (Rossum and Villarruz, 1961) and total carbohydrate (Yemm and Willis 1954). Chlorophyll estimation of leaves was done at regular intervals following Armon, 1954. To determine dry weight fractions individual plants were dug out separately with intact root system. They were washed, oven dried and weighed to obtain dry weight fractions. The values were divided by plant age to obtain NPP, expressed as gram plant^{-1} day^{-1}. All data were analysed statistically (SD and students' test, Pence and Sukhatme, 1985) and are presented in Tables 1-5.

Observations, Growth Parameters: The data for growth parameters are recorded in Table 1. Sulphurdioxde caused significant reduction in growth at all phases of development. In comparison to control, there was a decrease in the height of the plants exposed to SO_2. The height of 70 day old plant was decreased by 7.60 and 11.84% in 667 and 1334µgrn ^{3} SO_2, respectively. A slight recovery was recorded in the later phases of growth. There was also a decrease in the root and shoot length. The reduction in the root length was maximum in 70 day old plants, where it recorded a decrease of 8.14 and 13.42% in 667 and 1334µgm^{-3} SO_2, respectively. Further, the reduction in the root length was greater than the reduction in the shoot length. There was a reduction in the average number of tillers per plant due to exposure to SO_2. The maximum reduction was observed in 30 day old plants in both concentrations of SO_2. The number of leaves per plant also recorded a decrease at all phases of growth. The maximum reduction recorded in 70 day old plants was 10.09 and 21.10% in 667 and 1334 µgm^{-3} SO_2, respectively. Exposure to SO_2 also caused a decrease in the growth index over control. The decrease was higher in 1334 µgm^{-3} SO_2. The phytotoxicity percent was highest in 70 day old plants.

A decrease was observed in the dry weight fractions of S02 treated plants, Table–2. The dry weight of shoot and root of 50 day old plant was reduced by 13.22 and 14.22 and 17.79 and 25.80% in 667 and 1334µgm^{-3} SO_2, respectively. NPP also recorded a decrease due to SO_2 exposure, and the reduction of 20% was recorded in 50 day old plants at 1334 µgm^{-3} SO_2. Shoot and root ratio increased over control with increase in SO_2 concentration. But the proportionate increase was lower in later phases of growth.

Yield Parameters: Flower initiation was advanced by 5 to 8 days in plants exposed to 1334µgm^{-3} SO_2 but the effect of 667 µgm^{-3} SO_2 was insignificant. SO_2 also affected the yield, table 3. The number of grains per spike was decreased by 8.25 and 15.46% and harvest index by 5.65 and 9.80% in exposure of 667 and

1334µgm^{-3} SO_2, respectively. The weight of 1000 grains decreased by 3.31 and 6.71 % and spike density by 4.23 and 7.94% in 667 and 1334 µgm^{-3} SO_2, respectively.

Biochemical Changes: A reduction was observed in chlorophyll a, b and total chlorophyll contents of leaves of the plants exposed to SO_2. A maximum reduction of 11.61 and 18.18% in chlorophyll b was recorded in 70-day-old plants at the exposures of 667 and 1334µgm^{-3} SO_2, respectively. The ratio of chlorophyll a/b also decreased significantly and was directly correlated with the duration of exposure. A decrease was recorded in leaf extract pH of plants exposed to SO_2. Ascorbic acid content also decreased as a result of SO_2 exposure and decrease was 0.03 and 0.06 (mg/gfwt) in 667 and 1334 µgm^{-3} SO_2, respectively. Compared to control, the carbohydrate content of seed, stem and leaves were also adversely affected. Maximum reduction was in seed carbohydrates, followed by leaves and stem, Table 5. Exposure to SO_2 also increased sulphur accumulation in the leaves. The foliar sulphur content was recorded as 0.28, 0.34 and 0.41 mg/gdwt, in control and 667 and 1334µgm^{-3} SO_2, respectively. The total phosphorus content of leaves decreased significantly when the plants were exposed to S02. Foliar injury was observed after 15 days of exposure in 667µgm^{-3} SO_2, and after 40 days in 3341µgm^{-3} SO_2. The injury symptoms appeared in the interveinal regions at the tips and margins of the leaves in the form of chlorosis but later on they were converted into dark brownish nacrotic lesions. The extent of foliar injury increased with the increase in the concentration of SO_2 and duration of exposure. There was also a significant reduction in the leaf area due to SO_2 exposure.

Ameliorating Effects of Urea Spray on SO_2 Exposed Plants: The extent of foliar injury due to SO_2 pollution decreased when the plants were sprayed with urea. Plants exposed to 667 µgrn^{-3} SO_2 and sprayed with urea, did not show any foliar injury symptoms at all but in 1334 µgm^{-3} SO_2 the foliar injury was 11.40% when the leaves were sprayed with urea against 17.84% when no urea spray was done. In urea sprayed plants the reduction in leaves was also less than the unsprayed plants, Table 5. The reduction in plant height due to SO_2 exposure was much lower when the plants were sprayed with urea. In 70-day-old plants the reduction in plant height was only 4.32 and 6.14% in urea sprayed plants against 7.50 and 11.84% in unsprayed plants in exposures of 667 and 1334 μgm^{-3} SO_2, respectively. The urea spray had also similar effects on root and shoot length. The growth index was higher and phytotoxicity was lower in plants sprayed with urea. The average number of tillers and leaves per plant was also higher in urea sprayed plants than unsprayed plants but was lower than control. The reduction in fresh weight was also lower in urea sprayed plants than unsprayed plant, Table 1. There was a recovery in dry weight fractions of SO_2 exposed plants when sprayed with urea, Table 2. A similar response was also observed in NPP. There was also decrease in shoot root ratio of the plants sprayed with urea. Flowering was slightly advanced in plants exposed to SO_2 but the flowering time of the plants sprayed with urea was the same as that of control plants. A recovery in yield was also recorded in plants sprayed with urea but the yield was still less than the control. Urea sprayed plants also showed a recovery in chlorophyll a, b and total chlorophyll contents. The chlorophyll content of leaves of urea sprayed plants was higher than those of unsprayed ones; yet lower than the control plants. In 667 and 1334 μgm^{-3} SO_2, respectively, the decrease over control was 7.07 and 10.60% in chlorophyll a and 4.34 and 7.60% in chlorophyll b in urea sprayed plants and 11.61 and 18.18% in chlorophyll a and 9.78 and 14.13% in chlorophyll b in unsprayed plants. A similar trend was also observed in chlorophyll a/b ratio. Leaf extract pH and ascorbic acid content also recorded a recovery following the urea spray. In plants exposed to 667 and 1334 μgm^{-3} SO_2 and sprayed with urea, ascorbic acid contents recorded a decrease of 2.32 and 4.65% against in 6.97 and 13.95% in unsprayed plants. The reduction in carbohydrate contents of seeds, stem and leaves was lower in urea sprayed plants than unsprayed plants for instance in 667 and 1334 μgm^{-3} SO_2, respectively. The reduction (in comparison to control) in leaf carbohydrate was 6.42 and 10.17% in urea sprayed plants against 12.49 and 19.27% in unsprayed plants. Sulphur accumulation in leaves decreased with spray of urea. On the other hand, there was recovery in absorption rate of phosphorus in urea sprayed plants. Thus the present study indicates that urea supplies nitrogen to wheat plants which decrease their susceptibility to sulphurdioxide pollution.

Discussion: Foliar injury·symptoms are direct manifestations of phytotoxic nature of SO_2 (Hill et al., 1974) and is associated with the formation of toxic ions (H^+, HSO_3^{-1} and SO_3^{-1}) on the dissolution of water which caused local damage (Malhotra and Hocking, 1976). The exposure of *Triticum aestivum* to 667 and 1334 μgm^{-3} SO_2 caused foliar injury to the extent of 19.20 and 27.84% and leaf area reduction to the extent of 16.14 and 24.60%, respectively. The plants exposed to 1334 μgm^{-3} SO_2 and 5% urea spray, the foliar injury was only in the form of chlorotic patches on the leaf margins while the acute injury was also observed in plants which were not supplied with urea. With the cumulative dose of SO_2, the severity of leaf injury markedly increased in plants that were not supplied with urea. This suggests that urea neutralizes the acidity caused by SO_2 and thereby reducing the effects of SO_2 pollution. Guderian (1977) also reported that the use of fertilizers in the right proportion can help to counteract SO_2 damage to plants and is effective in reducing necrosis. In the present study, the growth of plants was adversely affected by SO_2 exposure. Data are in previous reports which indicate that SO_2 caused significant reduction in plant growth (Ashenden and Mansfield, 1972; Dell et al. 1979, Kumar and Singh 1985, Baker et al. 1986, Subhash and Singh 1989). The reduction in biomass may be due to low translocation as a result of reduced photosynthetic activity (Saxe, 1983). The plants subjected to SO_2 stress but sprayed with urea showed more healthy growth in terms of plant height, shoot and root length, number of leaves and tillers over the plants which were not supplied with urea. Urea supplies nitrogen to the plants which is utilized in protein synthesis. This was the reason of the comparatively healthy growth of plants sprayed with urea even under SO_2 stressed conditions. High sensitivity of chlorophyll a hampers the plant growth considerably as it plays a very vital role in the process of photosynthesis (Malhotra, 1977). The breakdown of chlorophyll may be attributed to SO_2 induced removal of Mg+ ions by two atoms of hydrogen from the chlorophyll molecules which convert chlorophyll to phaeophytin, changing the light spectrum of chlorophyll molecules (Rao and Leblanc 1966, Shimazaki et al. 1980). On the other hand, some workers are of opinion that SO_2 competes with CO_2 for the reaction site on RuBP carboxylase (Ziegler 1972, Mansfield and Jones 1984), the activity of RuBP carboxylase is thereby affected which in turn affects the photosynthetic rate and causes growth reductions. But when the plants were supplied with urea no marked change in chlorophyll content was observed due to SO_2 stress. Leaf extract pH and ascorbic acid content of *Triticum aestivum* recorded a recovery due to urea spray. For instance, ascorbic acid content recorded a decrease of only 2.32 and 4.65% against 6.97 and 13.97% in unsprayed plants. The reduction in carbohydrate content of seeds, stem and leaves was found comparatively much lower in urea sprayed plants than unsprayed plants. Killer and Schwager showed that continuous exposure of plants to SO_2 reduces ascorbic acid content long before the appearance of visible foliar injury symptoms, since ascorbic acid content acts as a strong reductant and is responsible for photoreduction of photo-chlorophyllid. Its decrease may lead to several physiological complications in plants (Rudolph and Dukatsh, 1966). According to Koziol and Jordan (1978) reduction in carbohydrate content in SO_2 pollution may be due to increased respiration and decreased CO_2 fixation. The foliar sulphur accumulation increased significantly due to SO_2 exposure but a marked decrease was observed in sulphur accumulation when plants were sprayed with urea. Stephen (1972) also observed that as a result of application of fertilizer there was 39% decrease in sulphur content of 1-year-old spruce needles. On the leaf surface urea combines with SO_2 and forms ammonium sulphate which prevents lowering of leaf pH (Thomas et al. 1944).

Phosphorus accumulation in plants exposed to SO_2 and sprayed with urea was higher than in plants exposed to SO_2 only. This is because urea spray reduces harmful effects of cumulative doses of sulphurdioxide. The SO_2 exposed plants flowered earlier than the control. The spike maturation was also advanced. Singh and Rao (1982) also reported an advanced flowering in *Cicer arietinum.* This may be due to the fact that under stressed conditions plants respond by rapidly completing their life cycle. There was significant reduction in yield (number of grains per spike, weight of 1000 grains and harvest index) due to SO_2 pollution. The reduction in yield as a result of SO_2 pollution has been reported in some other cereal crops, for example rice, wheat (Katz et al. 1984 and Kohet et al., 1987). The reduction in photosynthesis leads to decrease in weight of seeds, seeds per spike density, spike density enhances the reduction in total yield but when the plants were sprayed with urea yield was substantially recovered. Urea spray stimulates growth by inhibiting foliar injury and chlorophyll damage. The present study thus indicates that nitrogen supply to wheat plants decreases their susceptibility to SO_2 pollution.

REFERENCES

Agarwal, M., Nandi, P.K. and Rao, D.N. (1982), Effects of Ozone and Suphurdioxide Pollutants and in Mixture on Chlorophyll and Carotenoid Pigment of Oryza Satia, *Water Soil Air Pollution*, Volume 18, pp. 449-459.

Armon, D.L. (1949), Copper Enzymes in Isolated Chloroplasts, Polyphenol Oxidase in Beta Vulgaris, *Plant Physiol*, Volume-24, pp. 1-15.

Ashenden, T.W. and Mansfield, T.A. (1972), Influence of Wind Speed on the Sensitivity of Rya Grass to S02, *J Experimental Botany*, Volume 28, pp. 729-735.

Baker, C.K., Colls, J.J., Fullwood A.E. and Seaton, G.G.R. (1986), Depression of Growth and Yield in Winter Barley Exposed to Sulphurdioxide in the Field, New Phytol, Volume-104, pp. 233-241

Bell, J.N.B., Rutter, A.J. and Relton, J. (1979), Effects of Low Levels of Sulphurdioxide on Growth of Lolium Perenne, *New Phytol*, Volume 83, pp. 627-644

Guderian, R. (1977), Air Pollution: Phytotoxicity of Acid Gases and its Significance in Air Pollution Control. *Ecological Studies*, Volume 22. Springer Verlag, Berlin.

Hill, A.C., Hill, S., Lamb C. and Barrett T.W. (1974), Sensitivity of Native Desert Vegetation to SO_2 and NO_2 Combined, Air Pollution Control Association, Volume 29, pp. 153-157.

Jackson, M.L. (1958), Soil Chemical Analysis (Bombay, Asia Publication House).

Kats, M.G., Thompson C.R., Olszyk D.M. (1984), Effects of 03 and SO_2 on Annual Plant of Majave Desert, Air Pollution Control Association, Volume 34, pp. 1017-1022.

Killer, T. and Schwager, H. (1977), Air Pollution and Ascorbic Acid, *European Journal for Pathology*, Volume 7, pp. 355-380.

Kohut, R.J., Amundson, R.G., Laurence, J.A., Colavita, L., Pvanlcukch and King, P. (1987), Effects of Ozone and Sulphurdioxide on Yield of Winter Wheat, *Phytopathology*, Volume 77, pp. 741-774.

Koziol, M.J., and Jordan, C.F. (1978), Changes in Carbohydrate Levels in, Red Kidney Bean *(Phaseolus vulgaris L)* Exposed to Sulphurdioxide, *Experimental Botany*, Volume 29, pp. 1037-1043.

Malhotra, S.S. (1977), Effects of Aqueous Sulphurdioxide on Chlorophyll Destruction in Pinus Contorta, *New Phytol*, Volume 78, pp. 101-109.

Malhotra, S.S. and Hocking, D. (1976), Biochemical and Cytological Effects of Sulphurdioxide on Plant Metabolism, *New Phytol*, Volume 76, pp. 229-237.

Mansfield, T.A. and Jones, T. (1984), Growth Environment Interactions in SO_2 Responses of Grasses on SO_2 Pollution and Plant Productivity, (eds) W.E. Winner, H.A. Mooney and R. Goldstein (Stanford, Stanford University Press), pp. 3-26.

Pande, P.C. and Mansfield, T.A. (1985), Responses of Winter Barley to SO_2 and NO_2 Alone and in Combination, *Environmental Pollution*, Volume-39, pp. 281-291.

Panse, V.G. and Sukhatme, P.V. (1985), Statistical Methods for Agricultural Workers (New Delhi ICAR).

Rao, D.N. and Leblanc, F. (1966), Effects of Sulphurdioxide on the Lichen with Special Reference to Chlorophyll, *Bryologist*, Volume 69, pp. 69-72.

Rossum, J.R. and Villarruz, P. (1961), Suggested Methods for Turbidimetric Determination of Sulphate in Water, American Water Works Association, Volume 53, p. 873.

Rudolph, E. and Bukatsh, F. (1966), Die Chlorophyll (id) Unwant lung and ihre Bezichung zur photoxydation der Ascorbin saure bei elioliterten Keinpflanzen; *Planta*, Volume 69, pp. 124-134.

Saxe, H. (1983), Long Term Effects of Low Levels of SO_2 on Bean Plants *(Phaseolus vulgaris L)* 2nd, Immission Response Effects on Biomass Production Quantity and Quality, *Physiol Plant*, Volume 57, pp. 108-113.

Shimazaki, K., Sakaki, T. and Sugahara, K. (1980), Active Oxygen Participation in Chlorophyll Destruction and Lipid Peroxidation in SO_2 Fumigated Leaves of Spinach. In Studies on the Effects of Air Pollutants on Plants and Mechanisms of Phytotoxicity, Res. Rep Natl. Environ. Study, Volume 11, pp. 81-101.

Singh, M. and Rao, D.M. (1982), The Influence of Ozone and Sulphurdioxide on *Cicer acietenum L,* Indian Botanical Society, Volume-61, page 51-58.

Stefan, K. (1972), Results of Needle Analysis from a Fertilizer Trial in a Smoke Damaged Norway Spruce Stand, Mitt. *Forest Bundesvers*, Volume-97, pp. 52- 534.

Subhash, C. and Singh, V. (1989), Sensitivity of Triticale Hexaploide CV. Panda 6 to Sulphurdioxide. Proc. Indian Acad. Sci. (Plant Science), Volume 99, pp. 271-278.

West, A.W. and Gaeke, G.C. (1956), Fixation of Sulphurdioxide as Bisulphimercurate (11) and Subsequent Colorimetric Estimation, *Analytical Chemistry*, Volume 28, pp. 1816-1819.

Yemm, E.W. and Wills, A.J. (1954), The Estimation of Carbohydrate in Plant Extracts by Anthrone, *Biochemistry Journal*, Volume 57, pp. 508-514.

Zieglar, I. (1972), Effects of SO_2 on the Activity of Ribulose 1, 5-diphosphate Carboxylase in Isolated Spinach Chloroplasts, *Planta*, Volume 103, pp. 155-163.

Table 1. Effects of SO_2 and SO_2 + Urea on Shoot and Root Length, Number of Tillers and Leaves and Fresh Weight of *Triticm aestivum* cv. HDM 1553.

Parameter	*Plant age (days)*			*SO_2 (ugm^{-3})*		
		Control	*667*	*667 + U*	*1334*	*1334 + U*
Plant Height (cm)	30	40.82	39.29+	39.88+	39.08*	37.85+
		±2.89	±2.70	±2.74	±2.66	+2.55+
	50	66.94	62.51*	64.70+	60.56*	63.93+
		±3.88	±3.83	±3.79	±3.50	±3.61
	70	96.57	89.23*	92.29+	85.13**	90.64*
		±5.12	±5.04	±5.10	±4.92	±4.98
	90	109.60	104.33+	106.74+	100.74*	105.63+
		±6.22	±6.10	±6.18	±5.84	±5.96
	110	115.05	110.13+	111.50+	107.27*	109.76+
		±6.83	±6.57	±6.77	±6.34	±6.44
Shoot Length	30	31.14	30.14+	30.66+	29.22*	29.12+
		±2.15	±2.12	±2.10	±2.12	±2.08
	50	54.84	51.30*	52.88*	49.70*	52.48+
		±3.70	±3.62	±3.67	±3.58	±3.59
	70	81.22	75.13*	77.65+	71.84**	76.25*
		±3.91	±3.82	±3.83	±3.72	±3.76
	90	93.40	89.15+	91.10+	86.22*	90.32+
		±4.22	±4.15	±4.20	±4.13	±4.16
	110	26.61	24.80+	22.55*	22.55*	24.29*
		±4.22	±4.15	±4.20	±4.13	±4.16
Root Length (cm)	30	9.68	9.10*	9.22+	8.86*	8.73+
		±0.98	±0.92	±0.92	±0.90	±0.87
	50	12.10	11.21*	11.82+	10.86**	11.45*
		±1.25	±2.21	±1.26	±1.19	±1.20
	70	15.35	14.10*	14.64+	13.29**	14.39*
		±1.38	±1.36	±1.35	±1.36	±1.39
	90	16.20	15.18*	15.64+	14.52*	15.31+
		±1.45	±1.40	±1.42	±1.36	±1.39
	110	16.14	±15.33*	15.86+	14.72*	15.74+
		±1.48	±1.41	±1.45	±1.38	±1.42
Number of Tillers/Plant	30	4.80	4.30+	4.20+	3.70+	3.80+
		±0.03	±0.03	±0.03	±0.04	±0.03
	50	6.90	6.10*	6.30	5.80**	5.90*
		±0.07	±0.06	±0.07	±0.05	±0.05

	70	9.50 ±0.08	8.40* ±0.08	8.60* ±0.08	7.60** ±0.07	7.80* ±0.07
	90	–	–	–	–	–
	110	–	–	–	–	–
Number of leaves/Plant	30	4.10 ±0.05	4.00+ ±0.04	4.10+ ±0.05	3.70* ±0.04	3.80* ±0.04
	50	7.80 ±0.08	7.30* ±0.07	7.60+ ±0.08	6.70* ±0.06	7.40+ ±0.07
	70	10.90 ±0.14	9.80* ±0.12	10.40+ ±0.12	8.60** ±0.10	10.10* ±0.12
	90	12.50 ±0.17	11.60* ±0.16	12.10+ ±0.17	10.50* ±0.14	11.80+ ±0.15
	110	13.70 +0.20	12.50± +0.19	12.80+ +0.21	12.20+ +0.18	12.60+ +0.19
Fresh weight (g)	30	0.414 ±0.066	0.369+ ±0.005	0.394+ ±0.005	0.384* ±0.004	0.386* ±0.004
	50	7.80 ±0.08	7.30* ±0.07	7.60+ ±0.08	6.67* ±0.06	7.40+ ±0.07
	70	5.653 ±0.31	5.195* ±0.29	5.394+ ±0.30	4.986* ±0.28	5.322+ ±0.29
	90	7.131 ±0.39	6.547* ±0.33	6.918+ ±0.35	6.238* ±0.31	6.602 ±0.32
	100	7.952 ±0.40	7.112* ±0.39	7.503+ ±0.42	6.800* ±0.38	7.362* ±0.39

Values are in mean + SD; significance of difference from control, *P<0.05, **P<0.01, + non-significant

Table 2. Effects of SO_2 and SO_2^+ Urea on Dry Weight Fractions (g plants–1) and Net Primary Poductivity (g plant–1 day–2) of *Triticum aestivum* cv. HDM 1553.

Plant age (days)	*Part*	*Dry weight fractions*					*Net primary productivity (NPP) SO2 (µgm^{-3})*				
		Control	*667*	*667+U*	*1334*	*1334+U*	*Control*	*667*	*667+U*	*1334*	*1334+U*
30	Shoot	0.066	0.065^{+}	0.064^{+}	0.061^{+}	0.061+					
		±0.006	±0.006	±0.006	±0.006	±0.006					
	Root	0.017	0.015*	0.0015	0.014*	0.014*	0.003	0.003^{+}	0.003^{+}	0.003^{+}	0.003^{+}
		±0.001	±0.001	±0.001	±0.001	±0.001					
	Total	0.083	0.080	0.079^{+}	0.075*	0.075*					
		±0.008	±0.008	±0.008	±0.007	0.007					
50	Shoot	0.416	0.361*	0.375^{+}	0.356*	0.366*					
		±0.032	±0.028	±0.030	±0.024	±0.026					
	Root	0.062	0.051*	0.054^{+}	0.046**	0.052*	0.010	0.008**	0.009*	0.008**	0.008**
		±0.006	±0.006	±0.005	±0.005	±0.005					
	Total	0.478	0.412*	0.429*	0.402*	0.418*					
		±0.038	±0.031	±0.032	±0.031	±0.032					
70	Shoot	0.876	0.811^{+}	0.840^{+}	0.786*	0.835^{+}					
		±0.064	±0.058	±0.059	±0.052	±0.050					
	Root	0.339	0.304*	0.322^{+}	0.286*	0.317*	0.017	0.016^{+}	0.015^{+}	0.015^{+}	0.016+
		±0.022	±0.021	±0.023	±0.019	±0.020					
	Total	1.215	1.115*	1.162^{+}	1.072*	1.152^{+}					
		±0.092	±0.090	±0.089	±0.085	±0.088					
90	Shoot	1.189	1.056*	1.116^{+}	1.008*	1.062*					
		±0.010	±0.094	±0.096	±0.090	±0.091					

	Root	0.423 ±0.035	0.370* ±0.030	0.392+ ±0.033	0.348* ±0.028	0.379* ±0.031	0.017	0.016+	0.016+	0.015	0.016+
	Total	1.612 ±0.13	1.426* ±0.12	1.508+ ±0.10	1.356* ±0.098	1.441* ±0.10					
110	Shoot	1.250 ±0.12	1.161+ ±0.094	1.243+ ±0.096	1.089 ±0.090	1.20+ ±0.094					
	Root	0.449 ±0.38	0.402* ±0.036	0.428+ ±0.037	0.368* ±0.034	0.414* ±0.034	0.015	0.013*	0.015+	0.013*	0.014+
	Total	1.699 ±0.16	1.543* ±0.15	1.659+ ±0.16	1.462* ±0.14	1.621+ ±0.15					

Values are in mean ± SD; significance of difference from control, *P<0. **P<01. + non-significant.

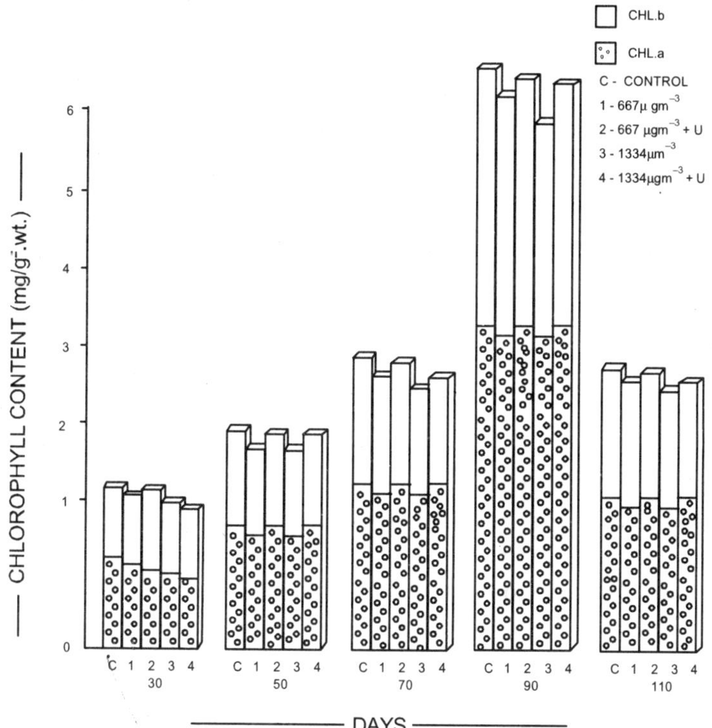

Fig. 1. Chlorophyll a, chlorophyll b and total chlorophyll content in 30, 50, 70, 90 and 110-day-old plants of triticum aestivum *cv. HDM 1553 exposed to 667, 667+U, 1334 and 1334 mg m^{-3} SO_2 $^{+U}$.*

Table 3. Effect of SO_2 on Carbohydrate Content (mg/gdwt) of Seeds, Stem and Leaves of *Triticum aestivum*

Plant			*SO2 (μgm⁻³)*		
Part	*Control*	*667*	*667+U*	*1334*	*1334+U*
Leaves	59.77	52.30*	55.93+	48.25**	53.69*
	±2.43	±2.38	±2.40	±2.36	±2.41
Stem	53.26*	48.53*	50.83+	44.70*	47.21+
	±2.42	±2.37	±2.40	±2.32	±2.35
Seeds	553.68	445.30*	493.52+	402.12**	485.66**
	±6.88	±5.64	±6.12	±5.20	±5.51

Values are in mean + SD; significance of difference from control, *P<0.05. **P<0,01, + non-significant.

Table 4. Effect of SO_2 and SO_2^+ Urea on Yeild Parameters and Harvest Index of *Triticum aestivum* cv. HDM 1553

	Treatment				
Parameter	*Control*	*667*	*667+U*	*1334*	*1334+U*
Grains/spike	47.20	43.30*	45.00+	39.90+	43.60+
	±2.00	±2.80	±2.80	±2.70	±2.80
Weight of	38.320	37.051+	37.513+	35.747*	36.559+
grains (g) 1000	±1.224	±1.201	±1.248	±1.307	±1.295
Spike density	1.89	1.81+	1.85+	1.74*	1.79+
	±0.09	±0.08	±0.08	±0.09	±0.08
Harvest index	55.25	52.13*	53.34+	49.83*	52.41+
	±2.46	±2.41	±2.39	±2.49	±2.45

Values are in mean ± SD; significance of difference from control, *P<0.05, + non-significant. U-Urea

Table 5. Effects of SO_2 and SO_2^+ Urea on Leaf Injury, Leaf Area, Leaf Extract pH and Ascorbic, Phosphorus, Sulphur and Carbohydrate Content of *Triticum* aetivum cv. HDM 1553.

	Treatment				
Parameter	*Control*	*667*	*667+U*	*1334*	*1334+U*
Leaf area injury (%)	–	19.20	–	27.84	11.40
Leaf area reduction (%)	–	16.14	10.64	24.60	13.76
Leaf extract pH	6.78	6.45	6.65	6.23	6.53
Ascorbic acid	0.43	0.40*	0.42+	0.37**	0.41*
(mg/g wt)	±0.03	±0.03	±0.03	±0.03	±0.03
Phosphorus	0.25	0.23*	0.24+	0.21**	0.23*
(mg/gdwt)	±0.02	±0.02	±0.02	±0.02	±0.02
Sulphur	0.28	0.34*	0.30+	0.41**	0.35*
(mg/100 mg d wt)	±0.03	±0.03	±0.04	±0.03	±0.03
Carbohydrate	59.77	52.30*	55.93+	48.25**	53.69*
(mg/d wt)	±2.43	±2.38	±2.40	±2.36	±2.41

Values are in mean ± SD; significance of difference from control, *P<0.05, **P<0.01, + non-significant. U-Urea

15

Environmental Movement in India

Kaushal Gupta and Poonam Dubey

Earth, the planet we inhabit, is a single, living, pulsating entity and the human race an interlocking extended family. The land, waters, and atmosphere which constitute its environment support some 0.36 million species of plants and more than a million species of animals. The unprecedented human interference into the environment has upset the delicate ecological balance of our planet. Ruthless exploitation of non-renewable natural resources has created havoc and if allowed to continue can result in a series of major ecological disasters that can disrupt life on this planet. Some of the most dramatic cases of tampering with the environment are man-made fires, slash and burn agriculture, mineral mining dumping of sewage and industrial wastes, the dangers of global warming and the attenuation of our ozone shield, the menace of deforestation leading to destruction of many species of flora and fauna, devastation of landscape by herbicides, introduction of pests into new areas, over-population, overgrazing coupled with drought, extensive air and water pollution, and the poisoning of the food chain, the malignant underworld of drugs and the alarming spread of communicable diseases. All these are problems which the human race shares in common, and in order to solve them people must try to attain as complete an understanding of their environment as possible.

The word "environment" can in itself be quite misleading since it encompasses everything that is around us. To the layman it means rather all things that can be damaged by man's activities. The real impact of man on the environment began when mining acquired a definite importance, that is only 40 to 50 centuries ago. And environmental effects took a significant importance only since the explosive industrial development began, around 1780, starting the exponential growth of world population and consumption.

The first environmental laws were enacted nearly at the same time. Napoleon ordered in 1810 that measures be taken by all factories to prevent damage or any kind of trouble to neighbours, crops, buildings, etc. Similar regulations were soon imposed in all industrial countries. Nevertheless it was only during the past thirty years that our world has really become "environment conscious". The wide-ranging character of pollution, of the irreversible damage it can cause in some cases, is a very recent stage of our conscious knowledge. And because it is still very young, the field of environmental protection remains incompletely defined in a number of domains.

There is no single environment movement in India. A large number of political initiatives, either single-issue-based or attempting to link a number of different issues – together with varying, and at times, competing ideologies – make up its mainstream. This needs to be understood from several different standpoints. One such standpoint would be to contextualize the development and significance of these movements with respect to larger politics. The second would be to classify and analyse the ideological trends in the movements as they have emerged in the last thirty years, to bring our political understanding up-to-date.

In India, the impact of 'modern development' was already felt in all its negative implications by the late 1960s and early 1970s. While Indian democracy was very proactive and reflected a large number of new concerns, this issue was never adopted by the major political parties. The ideology of development as the saving grace for the nation-state to acquire a new glory and recover one that was lost in the international community of nations was too strong a temptation for the Indian elite, especially, given the rich rewards it was able to reap for itself from it. Thus, not only did the educated 'middle class' not consider it in the planning process, but the political elite did not include it into the mainstream political agenda, too.

But the plight of the people was real and it was bound to have an expression. This was true for not only issues of the environment, but also a large number of other such issues, like those of access to land rights. So, in order to have their views heard, affected people, sometimes with the help of some conscientious educated people or even on their own, began to mobilise their voices and organise protests and have them communicated to the decision-makers and the powerful. These became known as the 'non-party political processes', 'the new social movements' and a number of different names – basically to set them apart, as being different from the standard kind of politics adopted by the political parties. The environment movement in India needs to be understood as part of this new development.

One other important characteristic of these movements was that they were not guided by any rigid political ideology. In the first place, had influenced people leading these movements. But many of them formulated their own understanding and interpretation of what they had read in terms of specific contexts, in which they were applying them and so came up with creative perspectives and explanations. But many of them were overwhelmingly influenced by Gandhian thought and action. That remains the case even today.

There were others that ranged from how development activity had created new and deepened the old structures of inequalities in the Indian society, on one hand, to those offering a critique of the development process itself, on the other. But whichever kind one might identify them with, in their focus and strategies of political action, it can be argued that they reflected the ground realities of the society and inculcated the values of radical and fringe politics, as practised in India for more than a century.

Some Environmental Movements in India

Several different kinds of issues have been the focus of these movements in India. Objects of development like dams, hydroelectric power projects, mining and mechanised fishing have been criticised on grounds that they have displaced people from their habitat and taken away livelihoods. Natural resources on which people are directly dependent – forests, land, seeds and water bodies such as rivers, lakes, the sea, groundwater and such others – have been defended by many movements from being over-exploited for fulfilling increasing demands of the market. And finally, many movements have focused on the fallouts of development – pollution (air, water, noise), slum settlements, liquor lobbies – of which, not all are directly related to the environment, but have serious implications for it. I will discuss some of these movements in detail.

Silent Valley Movement

The Silent Valley Movement in Kerala in the 1970s was one of the earliest and fiercest ecological controversies in post-independence India. The protests were against construction of hydel power project on the Kunthi river in the ecologically rich region called Silent Valley. The valley occupies only 8,950 hectares, but is surrounded by the Nilgiri, Nilambur and Attappadi forests that together comprise 40,000 hectares of pristine forest. There is a narrow gorge near the Mannarghat plains, which can be damned to form a reservoir to generate electricity. Technical investigations were carried out only in 1958. Preliminary works started in 1973, but due to shortage of funds, they were discontinued. At this time, no protests were heard. However, three years later, due to the intervention of some conservation-oriented bureaucrats, a task force was appointed by the National Committee on Environmental Planning and Co-ordination to go into the ecological problems of the Western Ghats. The task force submitted a report enlisting the disastrous consequences of such a project, which raised a storm. In January 1978, the Kerala Sastra Sahitya Parishad (a movement aimed at improving scientific temper amongst common people), organised the first mass signature campaign to stop the dam from being built. Nearly 600 academics, students and prominent citizens signed a memorandum addressed to the chief minister.

Chipko Movement

The Chipko Movement in Garhwal and Kumaon, in what is now Uttaranchal, also took place in the early 1970s, with people rising against the felling of trees for commercial reasons. The action that had fired the imagination of a whole generation was the hugging of the trees by the women of the villages, to protect them from the contractors' axes, undeterred by threats and eventually triumphant in regaining control over their forests. These two are perhaps the most well-known of the environmental movements in India.

Narmada Bachao Andolan (NBA)

One of the longest running battles against the construction of a dam has been that of the well-known Narmada Bachao Andolan, posited against the system of dams called the Sardar Sarovar Project, to be constructed on the river Narmada. The government's plan is to build 30 large, 135 medium and 3000 small dams to harness the waters of Narmada and its tributaries. The proponents of the dam claim that this plan would provide a large amount of water and electricity, which is desperately required for the purposes of development. Opponents of the dam question the basic assumptions of the Narmada Valley Development Plan and believe that its planning is unjust and iniquitous and the cost-benefit analysis is grossly inflated in favour of building the dams. Over the years, the extensive research conducted on behalf of the Andolan has demonstrated that the plans for the dams rest on untrue and unfounded assumptions of hydrology and seismicity of the area. Further, the construction is causing large-scale abuse of human rights and displacement of many poor and underprivileged communities. The critics also believe that water and energy can be provided to the people of the Narmada Valley, Gujarat and other regions through alternative technologies and planning processes that can be socially just, and economically and environmentally sustainable. Their research also shows how most rehabilitation schemes for people ousted with the building of other big dams have been inadequate. Thus, they have raised a range of issues, first of which is the necessity of such a big dam. They have repeatedly asked for lowering the height of the dam. The second issue is that of the resettlement package, without the proper distribution of which, they have threatened to continue their stay in the villages, where the release water would submerge them, thus resulting in mass suicide. Most importantly, however, they have questioned the model of development itself – arguing, instead, that an alternate model could be truly in the national interest. The NBA has by now become a kind of beacon for the environmental movements in India.

Mithini Village Movement

The village of Sonbhadr is rich in coal reserves. In 1989, the National Thermal Power Project (NTPC) first announced its plans to expand its operation into Mithini. The villagers were given small cash compensation and pressurised to move to a resettlement colony with the assurance that with further expansion, they would be given alternative livelihood options. This promise, however, remained unfulfilled even as displacement seemed imminent. In 1993, a petition was filed in the Allahabad High Court that gained a stay order against further construction until the promise of resettlement was fulfilled. Ignoring the directive, NTPC began to increase pressure on the villagers to leave their lands. In response to this, the movement against displacement took root in Mithini. Even as the villagers moved court to get the 1993 petition implemented, they faced the wrath of NTPC. They were verbally and physically threatened, paramilitary forces were used for intimidation, and the crops of protesting farmers were destroyed. In one instance, the only barrier that lay between the bulldozers and the village huts were the villagers, whole in front of the bulldozers, not allowing them to go further. However, the struggle took its toll. Two-thirds of the families left the village, unable to bear the aggression of NTPC and their own poverty. And only 18 families remained as signatories to the petition. Simultaneously, the struggle gained national and international support. Activists accused the World Bank, a major funding source for NTPC, of violating its own internal guidelines on settlement of displaced persons and demanded an internal inquiry. In one of the rare success stories in the history of people's movements, the panel found that the World Bank rules on resettlement were being violated and under pressure, the NTPC grudgingly give in. As a result of the struggle, they were able to secure a large compensation package.

Jharkhandi Organisation Against Radiation (JOAR)

The Uranium Corporation of India Limited (UCIL) has three operating mines in Jadugoda, Bhatin and Narwahpur villages of Jharkhand. In the 1960s, land was acquired in Jadugoda for constructing ponds, tailing ponds for dumping nuclear waste, and constructing residential colonies for the miners. Uranium ore is extracted, crushed and processed in Jadugoda and sent to the nuclear processing centre in Hyderabad for further purification into fuel rods. During the purification, depleted uranium mining tailings are discharged as waste, then brought to Jadugoda and dumped in large ponds in the midst of tribal villagers, or at times, on roads and even in fields. Radioactive waste from other parts of the country is also dumped in this area. Apart from causing environmental destruction, radiation and radioactive waste have severe health impacts, especially in terms of genetic mutation and slow death. The management of UCIL has stubbornly denied any harmful effects of radiation in Jadugoda and insists that radiation is 'within permissible limits'. The tribals, who work in these mines, are not given any protective devices like respirators, in the absence of which, they inhale dust and radon gas while they work. The initial mobilization of the tribals was around the issue of their resettlement and rehabilitation. The Jharkhandi Adivasi Visthapit Berozgar Sangh (JAVBS) was formed in 1989. Over the years, the people of Jadugoda have recognised the link between their health problems and radiation exposure. In 1998, the JAVBS was expanded to form Jharkhandi Organization Against Radiation (JOAR).

The JOAR has been involved in intensive campaigning in schools and colleges on effects of radiation on people and the environment. Soon after Pokhran II, it submitted a memorandum to the Government of India stating that the people of Jadugoda would not let anything that is destructive to human kind be extracted out of their land. In June 2002, the agitation of the people succeeded in preventing the acquisition of 15,000 acres of tribal land for open cast mining. The people of Jadugoda and JOAR are networking with other concerned organisations and peace movements for a nuclear-waste-free Jharkhand.

National Fishworkers Forum

With her 6000-km coastline and innumerable rivers, lagoons, lakes, reservoirs and ponds, India has one of the largest populations of 7 million fisher people in the world. One-third of them depend on marine fishing and the remaining two-third on fishing in a variety of inland water bodies. Despite being generally very poor and relegated to low castes, they have enjoyed a certain autonomy and dignity in the past.

With the introduction of export-oriented mechanized deep-sea fishing and aquaculture in the early 1960s, fishing economy underwent a marked technological polarisation. For the first time, fishing was seen as an industry, in which, anyone with capital could enter. Ignoring the skills and potential of a large number of fishing people, the government promoted technologies like bottom trawling, and so on, for large-scale harvesting of fish, which have depleted the marine resource base over the years. So, the fishing industry became polarised between the modern sector, able to make considerable profits from export, and a traditional sector confined to a domestic market with declining catches and fish stock. The 1970s witnessed violent clashes between the mechanized trawlers and fish workers, since they shared the same fishing grounds. The Majumdar Committee appointed to study the situation proposed the Marine Fishing Regulation Bill. The fish workers, on the other hand, started making new linkages and organising at local and state levels. This resulted in the formation of Trade Unions which came together to form the National Fish Workers Forum in 1979. Thus, it is an all-India organization that represents the interests of fish workers and unites these local movements.

Beej Bachao Andolan

Up in the hills of Uttarakhand is a place called Jardhargaon, where farmers are working hard to save the great variety of indigenous seeds that abound in different crops. Anticipating the possible imposition of monoculture practice by big transnational companies slowly entering the Indian market with hopes held out for higher yields through modified or laboratory tested seeds, the movement seeks to find ways of systematically saving the seeds of traditional varieties. They argue that these seeds have both quality and quantity yields, and are adapted to local growing conditions to boot, having been mutated naturally. Besides, they produce crops that people of a greater region now have a taste for and, if necessary, can be further mutated in the field for other qualities. Together with this, the andolan works to promote traditional agricultural practices, arrived at over generations of engaging with the terrain of those hills, and adapted to changing rainfall patterns and temperatures. Both of these are protective of certain natural resources and also of existing skills of local communities, recognized even by them as necessary for a comprehensive development process.

Environmental movements can be understood to be a part of what are known as 'new social movements' of the 1960s and the 1970s across the world. These movements came up for reasons specific to the politics of that time. During the Cold War, the two competing ideologies of Liberalism and Marxism had so completely dominated political discussion by their fixed categories of analysis, that some issues could not be included in them. These were issues of peace and disarmament, the models of development and their impact on environment, civil rights, gender, race and ethnicity. The political parties of most democratic countries seemed to be unaware of the importance of including these in their work agendas. As a result, people outside the political party processes, but concerned about these issues, mobilized others like themselves to raise their voice and call attention of their governments to them.

These movements, therefore, were usually single-issue-based and began as intuitive reactions to what seemed like harsh state action. To begin with, then, most of them had a social base in poor peasants or tribal people, as in Mithini or in Jadugoda. As they progressed, they found support among the urban middle-class intelligentsia of some radical political orientation, but mostly outside of the organized Left. This support was usually in terms of collecting and collating information – about laws, similar issues taken up elsewhere – and reaching out to a larger public by means of all media available. So, whenever any issue gained some kind of momentum, but needed to know whether there were any laws to support its demands, and whether and lawyers or judges would take up cudgels on its behalf, often urban students and professionals would garner support amongst themselves for it. For this, all kinds of information, about similar groups operating elsewhere and studies to support their arguments, would be brought together to be used for different purposes. Chief amongst these, as many movements and their urban supporters realised very quickly, was to interest and inform media professionals about this, enough for them to carry stories/features on these issues in their papers. Each one of these processes took a long time, because they were unfamiliar and unconventional. However, as a result of sustained political action on the ground, combined with inputs of information and support from outside, these movements took mature political positions, not only against the state or models of development, but also in favour of alternatives.

The unique position of environment movements makes such a connection possible. In all the movements the issue was one of the poor and their livelihood being taken away as a result of development, instead of the other way around. So, these movements bring to light one fundamental insight: *that development's benefits are for some and not always for all.* Hence, it can perpetuate or worsen inequality – social and economic – whereas its mandate and claim is to do the opposite. Yet, precisely on this ground, that of posing workable, road-based alternatives. It must be added though that there are a host of movements that are inspired and directed by urban middle-class positions on the environment, though many of them are in the category of non-governmental organizations, rather than movements.

For most of the last 30 years, these movements chose to remain outside the instiutionalized structures of politics. Disgusted by the corruption and cynicism of mainstream political actors, they did not participate in the political party/election process. They chose, instead, to challenge it from the outside, giving voice to concerns of people and making political accountability a reality, creating a base of legitimacy as the real representatives of the people. In most of the movements, the commitment of the leadership is to a very simple life, incorruptible in material terms, which is so blatantly evident and associated with mainstream politicians. Many of these people are not visible in the country's mainstream media, which never makes time and effort to find the genuine leader who still exists. And this is true of most new social movements, even other than those on the environment.

While most of these movements by themselves were strong, it was hard for them to sustain struggles over a long period of time with limited funds and people. At the same time, the state became more and more repressive in pursuing the new agenda of liberalization and privatization, implementing policies whether people wanted them or not. This was irrespective of the ideology of the party in power. With increased state repression, it became necessary to demonstrate 'people power', both in numbers and principle, and this required a new orientation.

In the last five years, however, there has been a different focus to the soul-searching of these movements. Relentlessly challenging the state, as they have done from the outside, they now feel the need to storm the bastion from within. So, it is believed by many that the legitimacy and standing that many of the leaders of these movements have, which is much more than elected politicians do, could be used more effectively if they are within the state. They could be trusted and would be accountable. Hence, there was a move to create a political front called the Lok Rajniti Manch (People's Political Front). This front did contest for some seats in the General Election of 2004, and expect to move towards a political party, drawing primarily from the constituents of the NAPM. While it is not clear what the outcome of their reflections will be, it can be expected to prove to be a creative intervention in politics.

Achievements

The principal achievements of the environmental movements in India are broadly of two kinds. The first is of introducing new rules and regulations that protected the environment and the interests of marginalized workers/tribals/ peasants. The second was that of introducing new ideas in the public space that did not exist earlier, making it possible to influence public policy and other interventions in the public space.

Criticisms

There are three kinds of criticisms that have been levelled against the environmental movement in India.

The first, most standard and powerful is from those who believe that environmentalists focus on the protection of nature *over and above* solving problems of poverty and underdevelopment, which naturally involve costs for some and benefits to others. The environmentalists go against a basic democratic principle that the benefits of the majority should be privileged because they prioritize the whims and wishes of small communities, reducing the significance of big projects to their petty needs. Thus, it is argued that they *do* put petty sectarian interests over the national interest. So, if some people are to be displaced by a dam and some benefit from the redirection of the water for irrigation and electricity, should not the latter be more important?

Second, the critics argue that traditional cost-benefit analysis that used to be conducted by economists to judge the profitability of any enterprise or project could now be amended and improved to be a social cost-benefit analysis, which could accommodate all the issues that the environmentalists raise.

Third, environmentalists have come under criticism from the dominant Left parties, especially in the early years. They accused these groups of narrowing the concerns of politics to the exclusion of the larger issues. So, questioning the environmental impacts of polluting or hazardous technologies was seen as antithetical to the interests of working class politics and the very issues of class, economic models and an overall improvement of the economic life of people. The critique they offered was as much about the issues the movements picked, as the fact that they mobilized political thought and action around single-point agendas. In focusing all political energy on this, the Left parties felt that larger concerns are either buried or postponed in the interest of the issue at hand, thereby deflecting attention away from broader social transformation.

However, these are concerns that these movements have consistently not only countered, but also taken serious cognizance of and used to reflect critically on their strategies and orientation. Consistently focusing on the poor and the disadvantaged, they have demonstrated that they wish the state and other agencies of development to approximate the definition of sustainable development, such that the environment is able to take the burden of development and the poor actually benefit from it. Thus, they have tried to demonstrate how they truly wish to work towards national interest, rather than the other way around. In practical terms, they have tried to substantiate this claim by offering models of and carrying out projects of alternative development that are people, environment and economic benefit-friendly.

REFERENCES

Agarwal, A. and S. Narain (1985). *The State of India's Environment: The Second Citizens' Report,* New Delhi: Centre for Science and Environment.

Sathi M.S. *Understanding Our Environment,* Commonwealth Publishers, New Delhi.

South Wick, C.H. (1996). *Ecology and the Quality of Our Environment,* William Grant Press, Boston.

Varshney, C.K. (1995). *Water Pollution and Management,* Wiley Eastern Limited, New Delhi.

D'Monte, D. (1985). *Temples or Tombs? Industry versus Environment : Three Controversies,* New Delhi: Centre for Science and Environment.

Gadgil, M. and R. Guha (1992). *The Fissured Land: An Ecological History of India.* Delhi: Oxford University Press.

Ministry of Commerce and Industry (2003). 'Govindarajan Committee on Investment Reforms'. New Delhi: Government of India.

Ministry of Environment and Forests (MoEF) (2004). *Draft National Policy on the Environment.* New Delhi: Government of India.

Rangarajan, M. (2001). *India's Wildlife History : An Introduction.* New Delhi: Permanent Black.

Sengupta, N. (1985). 'Irrigation: Traditional versus Modern', *Economic and Political Weekly,* Sepcial Number. August.

Whitcombe, E. (1993). 'The Costs of Irrigation in British India: Waterlogging, Salinity, and Malaria' in D. Arnold and R. Guha (eds.) *Nature, Culture and Imperialism: Essays in the Environmental History of South Asia.* Delhi: Oxford University, Press.

Sharma, M. *Landscapes and Lines: Environmental Dispatches on Rural India,* Oxford University Press, New Delhi.

16

Ecology vs Development Discourses in Post-Independence India: A Study of *Narmada Bachao Andolan* (NBA)

Sunil Kumar

"If the vast majority of our population is to be fed and clothed, then a balanced vision with our own priorities in place of the Western models is a must. There is no other way but to redefine 'modernity' and the goals of development, to widen it to a sustainable, just society based on harmonious, non-exploitative relationships between human beings and between people and nature."

Medha Patkar

Ecology and development as two important parameters of social science discourses have enlisted reactions and responses from different schools of thought at national and international arenas since the second half of the 20th century. The two terms started getting intermixing in wake of varying definitions and new dimensions thus changing the very nature of this relationship. Ecology conceived in terms of nature and development conceptualized in view of the economy brought out new meanings to the relationship of these two often contradictory but relatively complimentary phenomena in the global political reality.

Though the notions of ecology and development have had

their roots in Western discourses, the developing world started charting out their own independent paths in an attempt to reconcile ecological protection with sustainable development. Their failure to ensure an effective balance between ecology and development has, in fact, precipitated various social movements and new social movements in the developing world.

The present paper besides underlining changing discourses on ecology and development also attempts to unfold ecology-development paradox by analysing the role of one of the biggest and most sustaining ecological movements in global India, called *Narmada Bachao Andolan* (NBA).

Ecology and Environment: Key Postulates

The terms ecology and environment are usually applied interchangeably. Derived from the Greek word, ï6êïò, 'house' and -ëïãßá, 'study of'; ecology refers to the scientific study of the relations that living organisms have with respect to each other and their natural environment. Environment, on the other hand, is used to reflect natural life on earth with its focus on the totality of surrounding conditions.

Ecology is wider than environment for it covers various studies like physiology, evolutionary biology, genetics and ethnology – all to be encompassed under the broader notion of biodiversity. An understanding of how biodiversity affects ecological function is an important focus area in ecological studies which seek to explain 'life processes and adaptations, distribution and abundance of organisms, movement of materials and energy, development of ecosystems, and abundance and distribution of biodiversity in context of the environment' (www.wikipedia.com).

Viewed from this perspective the ecology movement is based on environmental protection, and is one of the several new social movements that emerged at the end of the 1960s. The growth of ecological movement was stimulated by a widespread acknowledgement of an ecological crisis of the planet. While the decade of the 1960s and 70s focused on the issue of nuclear weapons and nuclear power, the 80s revolved around the notion of acid rain.

Ozone depletion and deforestation figured prominently during the 1990s whereas climate change and global warming have now assumed great concern in the contemporary global society. The ecology movement has thus become an umbrella term highlighting various struggles and divergent protests undertaken by different groups, ideologies and attitudes across the globe in order to protect the planet.

Despite rooted in Western ecology, Indian ecology differs from its Western counterparts in the following way:

- Indian ecology focuses on survival, whereas Western ecology gives priority to the quality of life.
- Unlike India Western ecology supports collective action to achieve ecological goals.
- Indian ecology believes in direct action, whereas Western ecology supports lobbying, litigation, etc. to promote its ecological concerns.

Post-independence India witnessed ecological concerns substantially only with the emergence of various social movements like *Chipko*, *Aapikko*, and *Narmada Bacho Andolan* since the 1980s which kept questioning the primary developmental models centred round statism, centralized planning and the Westernized industrialism. Indian ecological movements in contemporary global India have tried to emphasize the developmental model with its minimal ecological destruction and maximal conservation of natural and material resources along with a focus on a reorganized social system enabling people to sustain their livelihood by enjoying the basic rights.

Development Discourses: Varying Turning Points

Discourses on development have failed to be conceptualized with definite yardsticks, common parameters, defined pathways, universal approaches and uniform strategies. Viewed as an organic process of change, the term 'development' has come to be viewed differently and interpreted variously by social scientists from the beginning. During 1940-1970 development came to be identified with 'employment' and 'basic needs'. The post-economic depression phase led economists like Keynes to argue that development could be possible only with the help of the State. This laid the foundation of **Bretton Woods System** in forms of International Monetary Fund (IMF) and the World Bank emerging as important financial channels assisting and orienting state economies towards development. This was the period when discourses on development started witnessing a change from 'redistribution from growth' (Hans Singer) into 'redistribution with growth' (Hollis Chenery) (quoted in Emmerij, 2006).

The decades of the late 1970s and 1980s constituted the **first turning point in development** which witnessed the emergence of Neo-Classical and Neo-Liberal School in social sciences. This period also set the stage for the birth of 'liberalization, privatization and globalization to be characterized by the acronym, LPG' (Sunil, 2009). Led by Margaret Thatcher and Ronald Reagan as the two heads of the leading world nations, namely, Britain and the United States respectively, this phase was characterized by a 'competitive open market system'. Development during this period got associated with 'growth' as against income and poverty. Defined in terms of 'Washington Consensus', the period stated that policy reforms would automatically ensure better living standards. It set the stage for structural reforms undertaken by the developed world individually as well as collectively which actually paved the way for the consolidation of economic blocks. The emergence of the European Union and ASEAN could be attributed to this development.

The decade of the 1990s constituted the **second turning point** in the history of development. It emerged with the popularization of the **Human Development Index** under the aegis of the United Nations Development Programme (UNDP). The period emphasized regional, cultural and nationalist variations in development. The focus of this period was to broaden the concept of development and narrow the range of development models, especially in the context of developing economies. Amartya Sen (2000) was the key champion of this development model, called the 'economic and social development model' which sought to bring on board political, cultural, social and human rights issues. Sen's economic and social developmental model emphasized tolerance and pluralism as the important principles of democratic procedures.

Amartya Sen's (2000) approach to development focused on the following two points:

- **Individual capabilities** – Amartya Sen argues that 'the poor are poor because their set of capabilities is small, not because what they do not have, but because of what they cannot do'. The list includes being able to lead a healthy and productive life, to communicate and participate in the community, to move about freely, etc.
- **Freedom** of choice as imbibed in the form of democracy. Amartya Sen argues that a 'country does not have to be deemed to be fit for democracy; rather, it has to become fit through democracy' (Sen, 1999). Democracy includes protection of liberties and freedoms, respect for legal entitlements, free discussions, etc.

The concept of development thus revolves around different turning points the focus of which has shifted from economic aspects to non-economic dimensions like social development, human development, ecological development and sustainable development. In case of India, while development has sought to address three significant concerns, viz., poverty eradication, health awareness and educational awakening; ecological protection and sustainable development have failed to gain ascendancy in the state's planning process and development discourses. Ecology and development as important planks of human sustenance have remained the very bases in the formation of various voluntary organizations, NGOs, civil society formations and resistance movements across India in the post-independence era.

Movements for Ecological Protection and Sustainable Development: Glimpses in Post-Independence India

Harmonizing ecology and development has always remained the focal point of debate in Indian society and polity for long. However, the mismatch between the two actually led to the emergence of various social movements in ancient, medieval and modern India. The most important of these movements could be traced to the *Chipko* Movement in northern India and *Appiko* Movement in southern India – both of which had their ancient lineages.[1] These two movements sought to preserve the ecology by resisting the state policies of rabid industrialization and unbridled commercialization of the natural habitat.

Beneath the Himalayan region of Garwal in the state of Uttarakhand, the ***Chipko* Movement**[2] made its unique place in the ecological history of post independence India. It was during the 1970s that by embracing the trees the people of the region resisted the State's attempts of massive forest cutting for commercial purposes. The development of the 'tree embracing' to be defined '*chipko*' in Hindi actually reflected an 'immense bonding between nature and human beings' (Biswal and Chand, 2008-09: 65). *Chipko* was not only the first ecological movement in post independence India but also constituted the first to have spearheaded the non-violent agitation for the protection of the nature.

The overwhelming success of the *Chipko* movement paved the way for various other ecological movements with protection of the natural habitat, especially the trees and forests, as their primary objective. Influenced by the success and guided by the momentum of the Himalayan movement another ecological movement tried to harmonize the relations between ecology and development in southern India. Described as **Appiko**[3], the movement began in the north *Kannad* region in the state of Karnataka in August 1983. The movement continued with great intensity for 38 days.

The Silent Valley movement in Kerala, *Chilka Bachao* Movement in Orissa, anti-Tehri dam movement in Uttar Pradesh, rain water harvesting movement in Rajasthan, Ganga Mukti movement, etc. were other similar ecological movements which tried to highlight the disjuncture between ecology and development in post independence India. All these movements according to Madhav Gadgil and Ramachandra Guha (1995: 385) have added a new dimension to Indian democracy and civil society besides posing 'an ideological challenge to the dominant notions of the meaning, content and patterns of development' (*ibid.*).

Characterizing these ecological movements as the social movements, Smithu Kothari (1995) argues that the strategy for most of thee social movements was 'to resist and restrict the exploitative character of the state and to seek to transform and democratize it in order to ensure its commitment to social justice and ecological sustainability' (Kothari, 1995: 426). India, according to Kothari, is a classic case of uneven development. 'Part of this unevenness is a consequence of the post-Independence emphasis on rapid industrialization, and part of it is a direct result of not addressing the endemic issues of social inequality. A Western-modelled industrialism has not only created and reinforced underdevelopment, but has led to a serious erosion of ecosystems that were the economic mainstays for a majority of people' (*Ibid.*432).

Sardar Sarovar Projects and the *Narmada* Movement: Social Concerns, Ecological Issues

No social movement had acquired such great prominence, wide intensity and massive participation for sustainable development and ecological protection in post independence India as were the *Sardar Sarvoar* Projects (SSP) and the *Narmada Sagar* Projects (NSP)[4] [to be called *Narmada* projects hence after] of the 1960s, both of which were 'the interconnected components of the large family of projects' (Ruitenbeek and Cartier, 1995: 2138) on the river *Narmada*. 'The *Narmada* river runs for 820 miles (1312 kms) through the Indian states of Madhya Pradesh, Maharashtra and Gujarat, passing through fertile plains and a series of hill ranges such as the Vindhyas and Satpuras' (Kala, 2001: 1995). It was believed that construction of a large number of dams on the river *Narmada* would ultimately result in the submergence of the entire *Narmada* valley.

The SSP and NSP paved the way for the *Narmada* protest movement from the 1980s onwards. The uniqueness of the *Narmada* movement lay in the fact that while the construction of dams on the river *Narmada* was opposed tooth and nail by the protesters in the states of Madhya Pradesh and Maharashtra in view of the submergence of their natural habitat and the loss of their very livelihood, the movement got support from the people and organizations[5] in the state of Gujarat as the *Narmada* dams were going to provide water to the war-ravaged regions of Kutch and Saurashtra of the state.

Constructing dams on major rivers of India was considered to be one of the fundamental policies of the scientific economic development under Nehruvian regime. The Nehruvian developmental model was to be built on the premise of strong infrastructural development. The Mahalanobis model as devised by Nehru focused on the critical application of the scientific and technological know-how for India's rapid and sustained economic development. Though the idea of 'damming the *Narmada*' was first mooted in 1946, the foundation stone on the river *Narmada* was laid by Nehru only in 1961. The *Narmada* projects were intended to provide irrigation, electricity generation and domestic water consumption benefits to approximately 40 million people inhabiting the states of Madhya Pradesh, Maharashtra and Gujarat.

While the *Narmada* projects came to be viewed as 'lifeline of Gujarat', they proved to be a death knell for the people of Madhya Pradesh and Maharashtra. Right from the beginning the projects had raised great controversies, the primary being the contradictory versions of development these projects were supposed to ensure for the people of the affected states. The Centre tried to sort out the problems by constituting the Khosla Committee[6] in 1965. The Committee, however, failed to solve the deadlock notwithstanding its overtures of bringing the affected states to the negotiating table for an amicable solution.

Exercising its powers under the Constitution regarding the inter-states water disputes, the Centre constituted an independent agency in 1969 called the *Narmada* Water Disputes Tribunal (NWDT) as a mechanism to solve the impending impasse among the three affected states. Partly because of political compulsions and partly due to administrative hindrances, it took almost a decade for the Tribunal to come out with its final award in 1979. Among many decisions of the award, the Tribunal allowed the construction of '30 major[7], 135 medium, and 3000 small dams' on the river *Narmada* across these three states in addition to raising the height of the major dams to 138.68 metres.

In view of a huge number of proposed dams on the river *Narmada*, an extensive canal of 450 kms comprising 31 branches as part of the long irrigation network, as well as a vast storage reservoir with a height of 455 feet, the *Narmada* projects came to be viewed as one of the biggest mega projects in the country. The Centre took some more time to initiate the formal beginning of the mega project. With an approval of financial assistance from the World Bank to a tune of US $ 450 million, the construction on the projects actually commenced in earnest from 1987 onwards.

With its objective of irrigating 50 lakh hectares of land, generating 3830 MW power capacity, and the target of benefiting 1.15 crore people of the affected states; the *Narmada* projects entered into the post-independence Indian history as a stupendous mega project bringing the ecology vs development debate to the centre stage of political discourse. While for the state planners the projects aimed at ensuring balanced economic development by fulfilling the basic needs of a vast majority of people, the critics termed the project for being 'the epitome of unsustainable development' (www.rightlivelihood.org) leading to various protest movements under the rubric of the *Narmada* protests.

William Fisher (1995) argued that the construction of dams on the river *Narmada* would submerge approximately 37,000 hectares of land in the affected states. It is believed that at least 100,000 people in 245 villages would be affected by the reservoir submergence. A large number of farmers, perhaps as many as 140,000 would lose their ancestral agricultural and residential land to the canal and irrigation systems to be built on the river *Narmada*.

The key arguments of the protests lay on the belief that construction of dams would actually result into forced displacement, a majority of whom would largely be the poor peasants, dalits, women and tribals. The construction of dams would also cause 'immense ecological damage through the inundation of forests, including prime habitats of rare species' (www.rightlivelihood.org).

Inadequate compensation, unsuitable resettlement and gross ecological damage were some of the concerns highlighted by the primary protest movements either resorted to by individuals alone or by their respective state organizations[8]. The protesters were also sceptical of the *Narmada* projects in view of the unexpected outcome of the past experience of large dams elsewhere in India highlighting the failure of the dams in terms of their projected benefits of hydropower, irrigation and drinking water to the respective displaced families.

The *Narmada* projects, in fact, generated a huge debate between the two opposing styles of development: one massively destructive of people and the environment in the quest for large-scale industrialization, whereas the other small-scale decentralized, democratic and ecologically sustainable development harmonizing the integration of the local with nature. Characterizing the current development approach as the prototype of 'gross distributive injustice' (Amte, 1990: 813), any effective developmental model would be successful if it involved the local people in a manner which would be 'non-destructive and sustainable both ecologically and financially' (*Ibid.* 818). The critics had advocated alternative models of development with a focus on 'energy and water strategy, based on improving dry farming technology, watershed development, small dams, lift schemes for irrigation and drinking water, and improved efficiency and utilization of existing dams' (www.rightlivelihood.org). Viewed thus the *Narmada* projects were going to become 'another human and ecological 'development tragedy' (*Ibid.*) in post-independence India.

Narmada Bachao Andolan (NBA): Critiquing the Nehruvian Developmental Model

Spearheading the ecology vs development debate by questioning the key developmental model in post-independence India was *Narmada Bachao Andolan* (NBA), a non-governmental organization set up by Medha Patkar and her colleagues in 1988. By leaving her doctoral studies in social work, Medha Patkar made several visits to the project sites, interacted with the affected people, approached public officials, addressed various meetings and encountered great revealing facts.

Though the protest movements against the *Narmada* projects had been going on for long, what Medha Patkar did was to bring all of them under the common umbrella of NBA in order to provide a united battle against the State and the faulty policy planners. The NBA increased both its organizational strength and physical presence by coalescing all the existing anti-dam protest organizations such as the Gujarat-based *ARCH-Vahini* (Action Research in Community Health and Development) and *Narmada Asargrastha Samiti* (Committee for people affected by the *Narmada* dam), Madhya Pradesh-based *Narmada Ghati Nav Nirman Samiti* (Committee for a new life in the *Narmada* Valley) and Maharashtra-based *Narmada Dharangrastha Samiti* (Committee for *Narmada* dam-affected people) who either believed in the need for fair rehabilitation plans for the people or who vehemently opposed dam construction despite a resettlement policy.

Medha Patkar was also able to bring various environmental and human rights groups into this united struggle against the state directed developmental project. The spirit of resistance arose from different quarters: 'the hamlets, villages, hills, forests and plains of the *Narmada* valley' (Kala, 2001: 2001). Turning the movement into a mega social movement highlighting the cause of social concerns, human rights issues and ecological degradation, Medha Patkar provided a great momentum both to the new social movements in general and ecological movement in particular in global India.

Under the collective endeavours and pioneering leadership of Medha Patkar, NBA raised significant objections to the ongoing *Narmada* projects by highlighting the following points:

- No prior clearance taken from the Ministry of Environment.
- Neither any information provided to the affected families nor their opinion taken.
- Provision of only short-term compensation rather than any long-term rehabilitation.

- Inadequate planning and implementation.
- Grossly unjust package of development

Unable to get any positive response from the State and substantial compliance to the grievances of the affected families, the NBA under Medha Patkar's leadership organized various *satyagrahas*, fasts, *dharnas*, protests and rallies garnering support of the movement across caste, class, community, state and region. Through her continuous and sustained agitations the NBA was able to highlight the plight of the affected families, mostly the poor farmers, tribals and *dalits*. The movement forced the state to carry out fresh environmental studies to ascertain the social and ecological damage to be caused in the aftermath of the projects. The problem of settlement of the displaced people called 'oustees' acquired significance in view of the introduction of new term called R&R [Rehabilitation and Resettlement]. The movement also tried to bring those people and their organizations; especially the *ARCH-Vahini* of Gujarat, to its fold by arguing that the 81% of the water starved *talukas* in Saurashtra would not get any water from the *Sardar Sarovar* Project.[9] Further, NBA claimed that two-thirds of Gujarat's drought-prone area would not be benefited by these projects.

The orientation of the *Narmada* movement as well as the reaction of the State to popular social agitation had various shifting stands. The State in its first instance started resorting to coercion rather than persuasion. It was finding it difficult to 'calculate and compare the projected costs and anticipated benefits' (Fisher, 1995). It had not anticipated any mass movement of this sort cutting across states and their boundaries. The initial focus of NBA was only to stop the dams per se, later on the focus shifted to stopping the construction of dams till the completion of the rehabilitation and resettlement work.

The NBA advocated peaceful methods of resistance comprising fasts, dharnas, and various mobilizing slogans. It began with its first slogan – '*Koi Nahi Hatega Bandh Nahi Banega* (no one will move, the dam will not be built) followed by *Doobenge Par Hatenge Nahin* (we will drown but we will not move)' (Patkar, 1995) and finally culminating into '*Amra Gaon Amra Raj*' (Our Village, Our Rule).

The movement had adopted the Gandhian strategy of non-cooperation since 1991 when the people of the areas didn't cooperate with the officials seeking a different type of information for relief and rehabilitation. At some stage the people didn't even allow the state officials to visit their areas for inspection and calculation.

One of the resounding successes of the NBA in its joint struggle against the construction of *Narmada* dams was directing the national movement towards the international arena. The continuous protests by NBA brought unprecedented success in terms of forcing the World Bank[10] to review its funding of the ongoing projects. Before the review commission could have taken any punitive action in its funding of the projects, India pulled out of the loan agreement with the World Bank in order to avoid international harassment. Finding great anomalies in the official reports and documentation, frauds in the environment compliance reports, massive corruption in rehabilitation work, NBA sought to seek the judicial shelter to counter the state and its detrimental developmental policies undertaken under the guise of *Narmada* dams in 1994.

Narmada Bachao Andolan and Judicial Verdicts

With its initial support from different sections of society and organizations fighting against the *Narmada* projects, NBA resorted to legal shelter in order to counter the apathetic attitude of the State towards the issues of displacement, rehabilitation and resettlement of the oustees. The NBA filed its first petition in the Supreme Court in 1994 by highlighting the plight of the oustees and requesting the Court for the immediate stoppage of work at different sites of the dams. Its success from the World Bank withdrawal and the eventual cancellation of the project in 1995 had further bolstered its confidence of enlisting justice from the apex court.

After deliberating the issue for six years the Apex Court gave its decision in 2000 by upholding the authority of the *Narmada* Water Disputes Tribunal, set up by the Centre, thereby allowing the construction subject to some reasonable conditions. The Court also upheld the constitution of the Inter State Administrative Authority known as *Narmada* Control Authority (NCA) for the purpose of securing compliance with and implementation of the decision and directions of the Tribunal. The Court stated that 'displacement of the tribals and other persons would not per se result in violation or their fundamental or other rights' (*Narmada Bachao Andolan* vs Union of India and Others [(2000 10 SCC 664]).

The Court in its judgment stated that 'every displaced family whose more than 25% of agricultural landholding is acquired, would be entitled to be allotted irrigable land of its choices to the extent of land acquired subject to the prescribed ceiling of the State concerned with a minimum of two hectares land' (*Ibid).* The Court further maintained that all the oustees, categorized as the Project Affected Families (PAFs), would be allotted a house/plot free of cost by the respective states. Moreover, the Court also ruled that the rehabilitation sites as provided by the three state governments had more and better amenities compared to the rural hamlets of the oustees. Moreover, according to the Apex Court, such moves would only lead to gradual assimilation of the oustees into the mainstream of society leading to their gradual betterment and progress. The Court fixed 'a time-frame so as to ensure relief and rehabilitation *pari passu* with the increase in the height of the dam' (*Ibid.*). In this connection the Court set up a Grievance Redressal Authorities (GRA) in each of the party states to monitor the progress of the dam.

The ruling of the Apex Court in 2000 assumed great significance in terms of its formalizing the implementation mechanism of the *Narmada* projects by emphasizing the initiative and role of the newly-constituted organizations like Rehabilitation and Resettlement Group, the Grievance Redressal Authority (GRA), the *Narmada* Control Authority (NCA) and other sub committees. The three-bench jury thus directed the states of Madhya Pradesh, Maharashtra and Gujarat 'to implement the award and give relief and rehabilitation to the oustees in terms of the packages offered by them' (*Ibid.*) by forcing the three state governments to comply with the directions of the Court.

The rulings by the Apex Court disappointed the NBA activists who were 'hoping for a more incisive examination of the rationale of the project and a greater sensitivity to the cry of the affected families' (Sachar, 2000: 5). In fact, the judicial verdicts created a predicament for the NBA. While it could not have defied the order of the Court by resorting to people's protests to stop the work on the *Narmada* projects, it was not entirely satisfied with the judicial ruling for conditional construction of the dams. Hence, the NBA once again decided to file a review petition before the Apex Court in 2000.

The final judgment of the Supreme Court came in 2005 wherein the Court ruled out any distinction between temporary and permanent oustees [project affected families] and stated categorically that raising of the dam[11] beyond its earlier approved 90 metres height should be based on the implementation of the relief and rehabilitation measures. The 2005 judgment of the Court cited three important considerations regarding the proposed benefits to the oustees, viz., 'major sons, choice of land, and extent of land' (*Narmada Bachao Andolan* vs Union of India and Others [(2005 10 SCC 328]). The Court stated that the sons of the PAFs who had become major one year before to the issuance of the notification for the land acquisition were entitled to be allotted land. Further, the distributed lands should not be rejected by the beneficiaries unless the lands were found to be non-irrigable or non-cultivable or unsuitable. Moreover, by upholding its earlier stand of 2000 judgment the Apex Court categorically asked the states of Madhya Pradesh, Maharashtra and Gujarat to ensure irrigable land to every displaced family, whose 25% or more agricultural landholding had been acquired.

The two important rulings of the Apex Court in 2000 and 2005 actually restrained the NBA from pursuing the path of struggle through further protests and marches. By dismissing the writ petition of NBA the Court in its final ruling had asked the organization to assume the role of vigilant observer in order to ensure a humane and painless resettlement work with due consideration to the ecological aspects of the projects. The judicial rulings which took nearly a decade did bring about a sense of perplexity by shattering the socio-political momentum of the *Narmada* movement. The

Narmada projects have now been going ahead without formal obstructions and hurdles by the peace marchers much to the dismay of the NBA activists.

Narmada Bachao Andolan: Critically Evaluating the Strategies and Goals

The *Narmada Bachao Andolan* differed from many other movements like the Naxalites, separatist insurgencies in the North East or secessionist movements of Punjab and Kashmir for being 'self consciously non-violent' (Kothari, 1995: 422). The movement which began with peaceful resistance in the beginning started resorting to violent and coercive techniques in the course of time.

The critics had argued that the NBA activists held up the project's completion, with its supporters attacking the local populace who had accepted compensation. Others have tried to equate *Narmada* protesters as 'environmental extremists' (www.wikipedia.com) who tried to halt the developmental projects through dubious unscientific facts. Instances could have been shown where the NBA activists had turned violent and attacked rehabilitation officers by causing damages to the official machineries. Charges were also levelled against the NBA activists for misleading the judiciary regarding the land entitlement claims of the beneficiaries.

The NBA faced many practical problems from its inception. Wide diversity, inherent complexity with 'floating constituencies' had hindered the very cohesiveness of the movement which also failed to bring all the protesters to a common platform of peaceful resistance. The state governments under political compulsions tried to wedge a great divide amongst various constituents of the movement. Diverse dialectics spoken by people in different areas of the affected states also created a stumbling block in the united struggle of the *Narmada* movement against the repressive state governments.

Despite these limitations the NBA is credited with unprecedented success in its struggle to highlight the ecology vs development debate by questioning the very foundation of the

Nehruvian developmental model. The movement succeeded in bringing the right to livelihood, right to habitat and right to control one's own life to the centre stage besides rights to consultation, information, participation and civil liberties. The movement was also successful in bringing great luminaries, social, legal and academic to its common platform. The names of Baba Amte[12], Arundhati Roy[13], Aamir Khan, Ali Kazimi[14], Anand Patwardhan[15] assumed great significance in pioneering the movement towards its logical conclusion.

Concluding Observations

While examining the political trajectory from colonial India to global India one encounters several significant trends in the developmental discourses in the country. The issue of ecology had always remained a pervasive feature of India's developmental history. The *Narmada Bachao Andolan* tried to project the ecology vs development debate in a broader canvas thus enlisting reactions and responses of the policy makers, planners, human rights activists, social scientists and the common populace both within and outside the Indian nation. It was because of the continued interventions, comprehensive policy influence and consistent mass movements undertaken under the umbrella of NBA that the issue of ecological development, rehabilitation and resettlement got its respectable place in the developmental discourses of the country.

Spearheading the people's struggle over the issue of sustainable development and ecological protection, the NBA has made path breaking contributions to the social and environmental cause both nationally and internationally. Its sustained struggles and consistent protests have largely facilitated the 'democratization of project planning and implementation at local, national and international levels by making it clear that society and actors have a crucial role to play in planning and governance' (Sen, 2000: 10).

Notwithstanding some limitations and shortcomings, the *Narmada Bachao Andolan* successfully highlighted the issues of development, displacement and rehabilitation by chalking out a multi-pronged strategy at the executive, legislative and judicial level on the one hand and championing the rights of affected people – farmers, labourers, tribals, dalits, women, fishermen and others on the other. By undertaking its sustained struggles for highlighting the ecology and development debate as well as by linking the same with the basic issues of people's development, NBA had charted out a unique place in the history of post-independence India.

NOTES

1. Both the *Chipko* and *Appiko* Movements were based on the protection of trees and forests which had their roots in the Vishnoi Movement in Rajasthan in 730 AD.
2. The centre of the movement was village Reni in district Chamoli in Uttarakhand. The Forest Department of the State Government leased out 2451 forest trees to a sports company of Allahabad in Uttar Pradesh, viz., Simond. Under the leadership of Gora Devi the women members along with the village men carried out the first ecological movement in post-independence India in March 1974 called the Chipko Movement. The movement got tremendous support from many social activists.
3. The centre of the *Appiko* Movement was Salgani and other adjoining villages in Northern Kannad region in the state of Karnataka. The movement spread to other tribal areas of Bengaon, Harsi, Nidgod, etc. The youth and women tried to salvage the forest trees with the same methods of the Chipko Movement by aligning themselves with the trees. For details, see Biswal and Chand, 2008-09.
4. The Sardar Sarovar Project (SSP) consists of a dam (Sardar Sarovar located about 100 kms east of Bharuch, Gujarat), an irrigation canal network, a riverbed powerhouse, a canal-head powerhouse and power transmission lines. The Narmada Sagar Project (NSP) comprises one large dam [Narmada Sagar, about 100 kms south-east of Indore in Madhya Pradesh] and two medium sized dams [Omkareshwar and Maheshwar]. For details, see H. Jack Ruitenbeek and Cynthia M. Cartier (1995). "Evaluation of *Narmada* Projects: An Ecological Economics Perspective", *Economic and Political Weekly*, Vol.XXX, No.34, 26 August.

5. ARCH-Vahini (Action Research in Community Health and Development) came out in support of the movement in the state of Gujarat.
6. The report of the Khosla Committee was accepted only by Gujarat, but rejected by the other two affected states, viz., Madhya Pradesh and Maharashtra.
7. Some of the large dams include Sardar Sarovar, *Narmada* Sagar, Indira Sagar.
8. Gujarat-based *Narmada* Asargrastha Samiti, Madhya Pradesh-based *Narmada* Ghati Nav Nirman Samiti, and Maharashtra-based *Narmada* Dharangrastha Samiti.
9. The governmental data shows that 56 out of the total 69 talukas of water starved Saurashtra, two-thirds of Gujarat's drought-prone or arid *talukas* and 42 out of the state's 47 tribal *talukas* and pockets are to get no water from the SSP. For details, see Baba Amte (1990). "*Narmada* Project - The Case Against and An Alternative Perspective", *Economic and Political Weekly*, Vol.XXV, No.16, April 21.
10. On the recommendation of Barber Coinable, the World Bank President, the Morse Commission was appointed to make an independent review of the *Narmada* projects. The Morse Commission in its independent review stated that 'performance under these projects has fallen short of what is called for under Bank policies and guidelines and the policies of the Government of India'. This resulted in the eventual cancellation of the projects by the World Bank in 1995.
11. The Supreme Court in its 2000 ruling had restricted the height of the major dams to 90 metres. Any further height [not more than the sanctioned total height of 138.68 metres] would be subject to rehabilitation and resettlement work on the part of the respective states.
12. Along with Medha Patkar, Baba Amte was the recipient of the 1991 Right Livelihood Award for their long and sustained struggle for the oustees under the auspices of the *Narmada* Bachao Andolan.
13. Winner of the prestigious Booker Prize for her book, *The God of Small Things*, Arundhati Roy became one of the strong forces behind NBA.
14. A noted film maker, Ali Kazimi, made a film titled, *Narmada: A Valley*

Rises documenting the five-week long struggle of the NBA undertaken in 1991. The film went on to win several awards and is considered by many to be a classic film on the issue.

15. A veteran documentary film maker, Anand Patwardhan, made an award-winning documentary on this issue, titled: *A Narmada Diary* in 1996.

17

Women and Environmental Degradation: The Role of MGNREGA in Managing Environmental Change

Sudhir Sharma, Alok Kumar and Shobha Sharma

Forests existed much before the existence of the human race on our planet. The ancestor of *homo sapiens* primarily depended for their subsistence upon forests and wildlife. Since the Vedic times, it has been known that nature and human kind (i.e. *Prakriti and Purush*) form an inseparable part of the life support system. This system has five elements: air, water, land, flora and fauna which are inter related, interdependent, co-evolved and co-adapted. The deterioration in any one of them inevitably affects the other four elements. If the deterioration is for a short-term, it repairs itself as it has enough resilience, but if it continues unabated, it adversely affects the whole system, and would ultimately start harming the existence of mankind on earth.

The whole system on which the existence of mankind depends is referred to as environment. Thus the term environment may be defined as, "The aggregate of all those things and set of conditions which directly or indirectly influences not only the life of organisms but also communities at a particular place".

The human environment is the earth we live on. It includes all the physical parts of the earth such as air, soil (minerals and rocks), water and all its living organisms such as animals and plants. The life-supporting environment of planet earth is called biosphere. The ecological factors which influence the existence of flora and fauna including man are as follows—

(i) Climatic Factors – Include all conditions related to aerial environment, i.e. light, temperature, humidity and wind.

(ii) Endorphic Factors – Relate to soil condition – Like temperature, soil constitution, nutrients, etc.

(iii) Physiographic Factors – Which are introduced by the structure, conformity or behaviour of earth's surface by topographic features like mountains, valleys, by geodynamic processes like erosion and silting.

(iv) Biotic Factors – Which include the influence of interaction of living organisms, i.e. of flora, fauna and man.

"Thus, environment is the sum total of all conditions and influences that affect the development and life of organisms".

The unlimited diversity of life form on earth, whether plants, or animals including man are most remarkable. If we trace the history of evolution, we find, man was one among many species, vying with all the others for survival. Then came his intelligence and scientific genius which led him to the top of the heap, where he started dominating nature, shaping the environment to suit him and using everything around him with scant regard for the natural interdependent ecological regimes. Man changed from fruit and root gatherer, the hunter and grower of food who worshipped nature to a more complex materialistic individual, who started destroying nature for the purpose of development.

The population started increasing and man began living in organized societies. The demand for food, shelter, clothes, etc. of the mounting population forced greater utilization of resources and development. Man, in order to meet his materialistic needs caused an unprecedented assault on nature and has mercilessly trampled upon our fragile eco-system, causing irreparable damage to the environment. Thus, the term development became synonymous with deforestation, and diversification and progress with pollution.

The adverse effect of environmental degradation can only be checked by bringing a transformation in the attitudes of the people from development-oriented strategy to the adoption of the concept of sustainable development. The change in attitude is necessary as it would fulfil the needs of both the present as well as future generations, and would assist in conserving and protecting the eco-system. Thus, sustainable development may be defined as-

Sustainable development is development that meets the needs of the present without compromising the ability of future generations to meet their needs.

Sustainable development contains two key concepts-

(i) The concept of 'needs' (essential) of the poor should be given priority.

(ii) The idea of limitations imposed by the state of technology and social organization on the environment's ability to meet present and future needs.

The main objective of development is the satisfaction of human needs and aspirations, which are limitless. The essential needs of vast populations in developing countries like India is food, clothing, shelter and jobs, which are not being met. Sustainable development aims at meeting these basic needs of all and extending to all an opportunity to satisfy their aspirations for a better life. But this does not mean that they start living beyond their ecological means.

The satisfaction of the needs of the people can be met by over exploiting resources, changing the direction of technological development, commercializing forests, diversion of water covers, etc. in the present day situation, but, all such methods will have disastrous effects on the environment especially to the future generation. This will threaten the life-support system and would go against the objective of sustainable development.

The exploitation of resources (realistic forces) can be done if they can regenerate. But, if they cannot be renewed like minerals, and fossil fuels, then their use will reduce the stock available for future generations. In such cases, the use of such resources should be carefully done so that they can last longer or as long as their substitute becomes available, i.e. the World Conservation strategy (a unique document on environment) which calls for a change in our attitudes from "***touch me not***" to "***use me wisely***".

Environment and development are two sides of the same coin. Sustainable development cannot take place by neglecting environment as it would one day close the door of development. That is why, the environment is to be protected from the needs of the poor, the greed of the rich and the careless application of technology. The excess protection of environment will put economic development at stake, which in turn would lead to poverty, starvation, backwardness and mis-utilization of resources, thus causing economic hardship. That is why, a judicious coordination between environment and sustainable development is necessary. This co-ordination is to be achieved not only within the country but also between the countries, as environment is not a state issue but a global one.

Women and Environment

Physical impacts resulting from accumulation such as rising temperature, rising sea levels, extreme events dramatically alter the natural balance of local and global ecosystems and infringe on human settlements. Consequently, vulnerable groups such as the poor, **especially women** face problems such as food insecurity, loss of livelihood, hardships due to environmental degradation which also lead to displacement and a whole host of potentially devastating economic and social consequences. It is the poor women who are vulnerable and bear the adaptation burden despite their insignificant contribution in environment degradation. Of course, in the rural areas women are much involved in agriculture, fishing, and other livelihood activities as men are, however, their roles and rewards differ. If gender roles in real life are clearly differentiated then each should be assisted to cope with the problems they face.

In 1975, the International Women's Year, which was declared in honour of the 25th anniversary of the United Nations Commission on the Status of Women highlighted women's environmental concerns for the first time (Tinker and Jaquette, 1987). The "Decade of Women" also began at this time. At the end of the decade, the United Nations Environment Programme (UNEP) sponsored a major conference in Nairobi, Kenya. One outcome of the conference was a synthesis of policies to advance women, the *Forward Looking Strategies to the Year 2000.* Many of these policies were subsequently adopted into Agenda 21 at UNCED as Chapter 24, *Global Action for Women Towards Sustainable and Equitable Development.*

It is widely acknowledged that the negative effects of environment are likely to hit the poorest people in the poorest countries the hardest, in other words: the poor are most vulnerable. Women form a disproportionate share of the poor in developing countries especially in communities that are highly dependent on local natural resources: Their problems should be addressed adequately. The paper is an analysis of the impact of climate change on women in the states of Jharkhand, Chhattisgarh and Madhya Pradesh and how MGNREGA has been effective in empowering the local women to manage the same.

Women and Climate Change Consequences

Impact on Livelihood

Women are more dependent for their livelihood on natural resources that are threatened by climate change. For instance, climate change causes a rise in the sea level, affecting the fishing community (both men and women) not only in terms of fish catch but also with regard to water scarcity, as sea water gets into fresh water. Besides, when the land is inundated, infrastructure (roads and houses) are damaged. Large scale migration from inundated areas is expected and much of the burden of migration falls on women.

As primary caregivers, women may see their responsibilities increase as family members suffer increased illness due to exposure to vector borne diseases such as malaria, water borne diseases such as cholera and increase in heart stress mortality

"Poor people are more vulnerable to climate change due to their limited adaptive capacities to a changing environment. Among them, the rural poor, and rural women and girls are the ones most immediately affected. Climate change impacts are not gender neutral". Poverty reduction through an integrated approach to local resource management is the key to development. There is a need to see gender issues in a broader perspective since they are still seen within rigid and restrictive lenses.

The author observes that gender is a significant dimension to take into account when understanding environmental change. It is important to understand how men and women may be affected differently by climate change considering their roles and responsibilities in the society. The knowledge thus gained would help improve actions taken to reduce vulnerability and combat climate change in the developing world. For example, perspectives, responses and impacts related to disaster events are different for men and women, as men and women have different social responsibilities, vulnerabilities, capabilities and opportunities for adjustment and unequal assets and power relations. Their physical abilities and the way they experience environmental changes and disaster can be different.

Rural women in developing countries are still largely responsible for securing food, water, and energy for cooking and heating. Drought, deforestation, and erratic rainfall cause women to work harder to secure these resources. So far, climate change has largely been conceived as a scientific process and less in social realms

We attempt to make a framework for such socio economic analysis. The introduction of the MGNREGA has helped in the creation of the assets and also reduced the impact of climate change partially.

Natural Resources Management

Women's role in natural resource management is considerable whether it is water, agriculture, forest or wildlife ecosystem, and coastal zones. Climate change will have an impact on the biomass which is the vital source for food, fuel, fibre, construction material and livelihood activities. In the states surveyed there is an increase in the forest cover, agricultural output and also increase in the water bodies. This has been possible as women have got both employment and opportunity to work in their own villages by the introduction of the MGNREGA.

Agriculture and Food Security

Climate change could mean extra hardship for farming activities, often carried out by women, especially in the villages, e.g. in paddy cultivation, cash crops such as cotton and tea plantations and so on. Women's contribution and participation can help or hinder environmental resource management. High dependency on agriculture, forest sectors and bio fuels could increase vulnerability and heighten the risk of environmental depletion.

For instance, agriculture will be seriously affected as developing countries, largely characterized by their vulnerability, weak institutional capacity and precarious financial situation, try to grapple with the problems of climate change. In India in the long term, small increases in temperature are, in aggregate expected to reduce crop yields and the area of arable land to a greater extent than other regions.

Women's active involvement in agriculture and their dependence on biomass energy would mean effective environmental management. The need to diversify energy resources and facilitate the introduction of substitution fuels for household energy consumption could well constitute the essential part of adaptation strategies.

Small farms are generally owned and operated by the poor, often women, who use locally hired labour, and distribute income within nearby locales, creating multipliers. These small farms have some advantages over the large farms in certain types of transaction costs: the supervision of labour, local knowledge, and food purchase for self consumption and risk reduction. But, if climate change also means that there is less agricultural land available, and the area of low potential land is increased, the need to increase land productivity to stimulate agricultural growth becomes all the more pressing. With changes in climate, traditional food sources become more unpredictable and scarce. This exposes women to loss of harvests, often their sole sources of food and income. With cash crops becoming scarce, food prices increase and make the situation even worse.

Water and Other Resource Shortages

Climate changes have changed existing sources of water. Women, largely responsible for water collection in their communities, are more sensitive to the changes in seasons and climatic conditions that affect water quantity and accessibility which makes its collection even more time consuming. Increased frequency of droughts means that women are walking greater distances to collect water, often ranging from 10 to 15 kms a day. This confronts women with personal security risks, keeps young girls out of school and imposes an immense physical burden—a plastic container filled with 20 litres of water weights around 20 kgs. Strengthening water resources and delivering systems will be done best with women's help and involvement. A survey of the villages showed that several available water sources had dried due to the drought and severe water shortage was seen. The MGNREGA came as a boon and there was a drive to de-silt and rejuvenate all the ponds and also dig new ponds in the villages ensuring a water source throughout the year.

Women's specific knowledge in soil conservation is now recognized by planners which incorporate low cost techniques adapted to local conditions, thus, women's role in adaptation programmes for water management is necessary to ensure effective delivery.

Climate change may exacerbate existing shortages of water. Women are largely responsible for water collection in their communities and are therefore more affected when the quantity of water and/or its accessibility changes

Forests

Forests are vital for food resources. Their unsustainable use would result in a shortage of non-timber forest products (NTFP) – which could lead to malnutrition and infant mortality. An increasing number of women depend upon forest resources as a major source of their livelihood in rural India. Forest products also serve as a source of nutritional and food supplements thus providing alternative nutrients, minerals and vitamins to the usual staple food. The tribal women have been engaged as workers in the states of Jharkhand

and Chhattisgarh as van poshaks and they are now looking forward to the benefits of the increased forest cover. The senior women in the village have been consulted and an effort has been made to plant trees of both medicinal and cash value. Thus the impact of deforestation will be reduced in the future. The increase in the forest cover will definitely bring a change in the local climate.

Preservation of non-timber forest products with women's help will ensure livelihood and food security. Women's strong role in preservation is remembered by the "***chipko movement***" in the 1970s, where women hugged the trees to prevent them being cut by timber companies. The National and the State Governments have to build embankments to protect the population and infrastructure in danger. In order to commit this task the government could use women's labour force. Women can be protected and also help protect themselves.

Health

Environmental change has affected health in a variety of ways, including:

- Increased spread of vector and water-borne diseases
- Reduced drinking water availability
- Food insecurity due to reduced agricultural production in some regions and hence malnutrition or disturbance of traditionally balanced diets. Women are more prone to malnutrition as they are the last to eat after feeding the family. As primary caregivers in many families, women may see their responsibilities increase as family members suffer increased illness. Further, in the developing world, women often have less access particularly in areas of environmental degradation, and where basic public infrastructure, including sanitation and hygiene, is lacking. Women are more in contact with water; they are more likely to be impacted.
- Increases in ground level ozone concentration could increase respiratory and cardiovascular morbidity and mortality. This is in addition to indoor pollution that women suffer which also leads to the same problems.

- Increases in mean temperature could facilitate the spread of malaria and dengue fever along the current edges of their geographic distribution in some regions, and increase the length of the transmission season for malaria, although the magnitude of the effect is thought to be smaller than previously estimated.
- Increases in daily temperature will increase numbers of cases of food poisoning in temperate regions.
- There are important prerequisites for adaptation that are currently not met in many parts of the world, e.g. access to primary health care and basic education.
- There has been progress in the design and implementation of climate health warning systems, established to reduce effects of weather extremes as well as for the seasonal prediction of infectious diseases. Limited evidence suggests that such systems can be effective, provided women are also introduced to this and are aware.
- Women are more likely to suffer heat stress and account for heat-related deaths, perhaps due to biological reasons.
- High precipitation tests the integrity of water management systems and increases the risk of outbreaks of water-borne diseases.

Additionally, they also observe that climate change has many gender specific characteristics:

1) Women are affected differently, and more severely, by climate change and natural disasters because of social roles, discrimination, poverty and intra household inequity,

2) Women are still underrepresented in decision-making about climate change, greenhouse gas emissions and adaptation/mitigation,

3) There are gender biases in carbon emissions. They should be included not only because they are most vulnerable but also because they have different perspectives and expertise to contribute.

Study of Climate Change—Its Impact and Solutions in Jharkhand

Jharkhand, is a state in eastern India, it was carved out of the southern part of the State of Bihar on November 15, 2000.

Jharkhand shares its border with the states of Bihar to the north, Uttar Pradesh and Chhattisgarh to the west, Orissa to the south, and West Bengal to the east. It is spread over an area of 28,833 sq miles (74,677 kms^2). Jharkhand has 24 districts, 211 blocks and 32.620 villages out of which only 45% are electrified while only 8484 are connected by roads.

According to census 2001, Jharkhand has a population of 26.93 million people, consisting of 13.88 million males and 13.08 million females. The sex ratio is 941 females per thousand males. The population consists of 28% of tribals, 12% schedule castes and 60% others. The density of population is 338 persons per sq km, among the districts, Gumla (148 persons per sq km) is the least densely populated while Dhanbad (1167 persons per sq. km.) is the most densely populated district.

The per capita income of Jharkhand is Rs. 21465/- at current prices (2007-08) and the population living below the poverty line is 45.3% of which 51.6% is in rural areas and 23.8% in urban areas. The literacy rate in Jharkhand is 53.56% of which male literacy is 67.30% and female literacy is 38.87%.

Observations in Jharkhand

Mahatma Gandhi NREGA has directly and indirectly improved the situation of the natural resources in the state of Jharkhand. Gradually, the women have started feeling that they are also part of the main stream and can contribute to the development of durable assets.

The attitude of women varies in the different regions of Jharkhand. Some areas like Pakur reflect a higher degree of empowerment in contrast to areas of Deoghar, Khunti, and Dumka.

The women are also taking part in the process of planning and decision taking. (Priority decided by tribal women of Ghagarjani, Hiranpur block for the development of water body).

Mahatma Gandhi NREGA has brought an improvement in the status of women. They work with dignity as their relation is not that of master and servant but is now of employee and employer under Mahatma Gandhi NREGA.

The women now present a picture of happiness and satisfaction and are thankful to government for starting the Mahatma Gandhi NREGA works. They are also increasing the number of days of employment.

Mahatma Gandhi NREGA works have generated water facilities at the door step for which women had to go miles. The carrying of water for long distances, up steep terrains, etc. caused severe health problems.

The Mahatma Gandhi NREGA has trained the women to share and care for the water bodies created. They have a strong feeling that now they are able to save time which they can use for social interaction and for caring of their families.

Details of the Works Done in Jharkhand

Districts	*Rural Connectivity*	*Flood Control and Protection*	*Water Conservation and Water Harvesting*	*Micro Irrigation Works*	*Provision of Irrigation Facilities to Land Owned by*	*Land Development*
Chatra	2444	1	8738	7	2	8
Dhanbad	1451	35	4827	120	393	244
Dumka	2678	3	7775	5	3284	2116
Garhwa	1880	1	6190	0	0	61
Giridih	3093	1	5036	45	35	333
Godda	1500	7	5140	0	0	12
Gumla	3119	92	13239	7	104	687
Hazaribagh	1741	8	3162	112	52	147
Koderma	733	1	2469	29	3	349
Latehar	3185	5	6895	0	0	158
Lohardaga	1312	1	4323	78	211	618
Pakur	1356	19	2385	0	0	65
Palamu	2171	5	3438	136	652	20
Ranchi	2017	1	9368	13	548	277
Sahebganj	2422	128	1178	321	265	229
Saraikela Kharsawan	1699	161	4197	2	1447	3342
Simdega	6703	3	7870	5	638	2819
West Singhbhum	4899	15	4911	284	607	87

Deoghar	2206	14	1944	37	1061	3964
East Singh-bum	2793	32	3679	121	276	682
Khunti	825	0	1814	33	308	39
Ramgarh	1087	0	2334	2	0	132
Total	**53940**	**542**	**117696**	**1368**	**10369**	**18121**

Source: Ministry of Rural Development, Government of India

Rural Connectivity

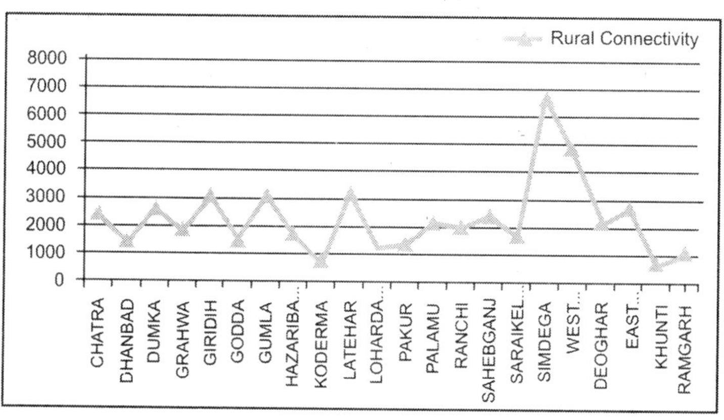

Flood Control and Protection

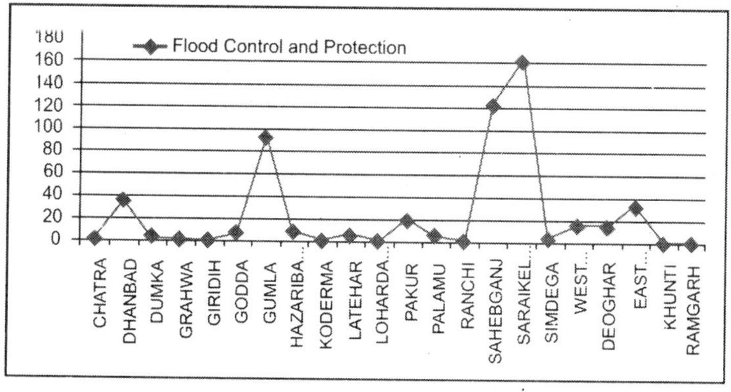

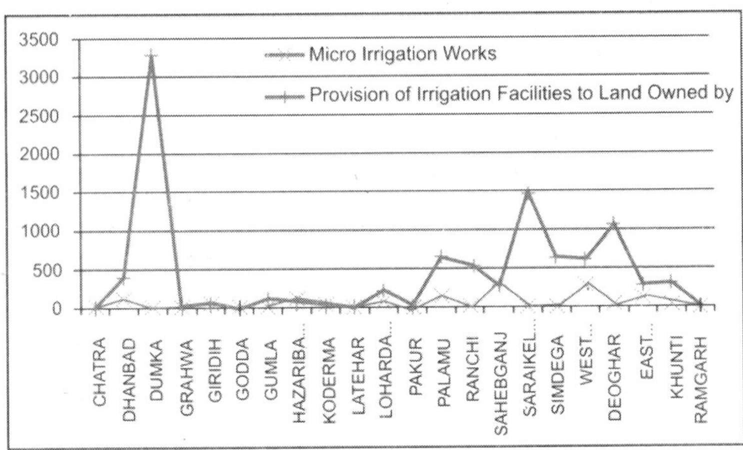

Water Conservation and Water Harvesting

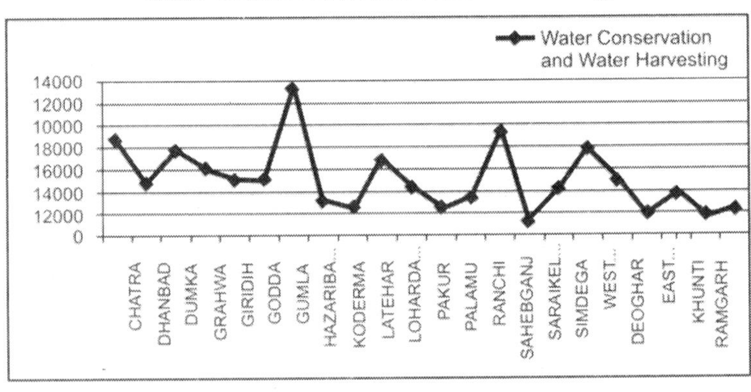

Land Development

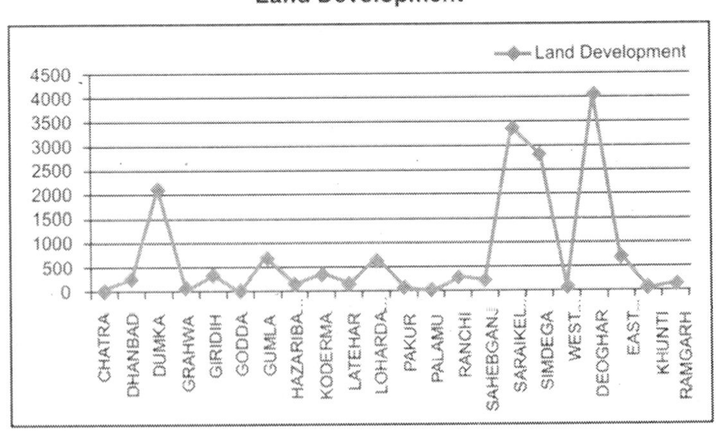

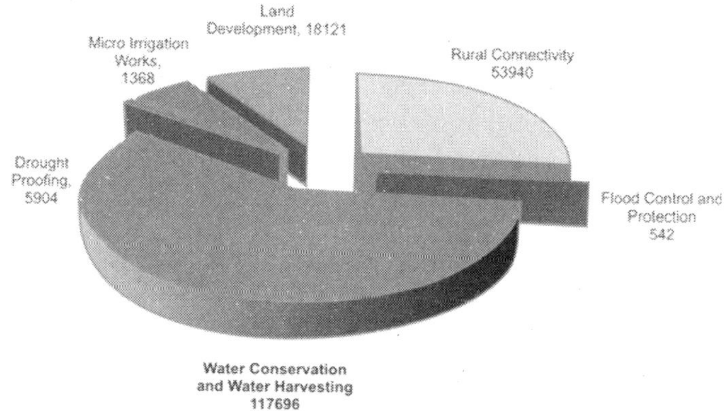

Way Forward

Since climate change is a long term issue, we need to focus on long term goals first and then see how to get there step by step through medium and short term measures. For our purpose we define short term as 1 to 5 years, medium term 5 to 15 years and long term beyond 15 years. However, even medium and long term strategies will have to start their trajectories **so that when the 5th year comes**, the preparatory ground is ready.

Better understanding, knowledge and information is needed and the approach should be flexible to steer as new information and knowledge open up.

Short Term Strategies

- Budget allocation: This is the best way to ensure beneficiaries get the required attention. The budgets are needed for preparation of the medium term capacity building and R&D and demonstration project.
- Task force to upgrade rural infrastructures: Municipalities need technical and financial help to upgrade infrastructure and to build capacity to deal with frequent floods, water shortages and other events as to help vulnerable population.
- Coping strategies for livelihood: Fishermen, farmers and poor population in coastal, mountain and arid zones will

have to be trained to adapt to changing climate and continue to make a living.

- Risk and insurance: While India is asked to share the burden of mitigation, counter gesture on sharing adaptation burden through global insurance for poor facing disasters and distress is in order. Perhaps a protocol to address this need is to be initiated. Such a protocol, if developed, should give special assistance to women.

Capacity building, knowledge management and participation:

- Prepare training module for training NGOs and private sector for various ecosystems and manifestation on climate change. These should not be too detailed to start with, but give broad hints and understanding for evolving programmes with consensus.
- Impart pilot training to a few countries and redesign modules and manuals from the ground level feedback. Replicate training on a larger scale.
- These are needed for adaptation as well as mitigation. Women's knowledge in adaptation could be used as a resource and needs to be documented. Often, this knowledge may be community specific.
- Women's participation is imperative if efforts to combat global warming are to succeed without their participation, many programmes may not succeed. We expect that the households with higher social capital to have higher probability of participating in dealing with climate change. Women's inclusion in decision making will ensure success and may reduce costs. Their insights may be not just valuable but essential.
- Women's participation in decision making is important to handle the issues in new perspectives of governance. Societies with greater role of women in energy planning and decision making one likely to formulate better village level or cluster level plans. The proportion of budget women received can measure women's empowerment and effective participation.

Achieving and Accelerating MDGs

Literacy and empowerment include development of safety nets, social and physical infrastructures and poverty eradication. These could be the best response strategy and efforts in that diversion here to be accelerated .They will bridge the gender gaps and eventually reduce gender differences.

Actions Needed

The following specific action is required to tackle the negative environmental impacts:

- Capacity building for new and gender specific programmes for adaptive farm practices.
- Integrated biomass strategies for food, fuel, fodder, and other basic needs including income generation.
- Investment in irrigation and rain harvesting schemes.
- Capacity building for waste management.
- Practices to prevent the adoption of soilless and drought resistant varieties.

BIBLIOGRAPHY

1. Agnes R. Quisumbing, Lynn R. Brown, Lawrence Haddad, and Ruth Meinzen-Dick, The Importance of Gender Issues for Environmentally and Socially Sustainable Rural Development in Agriculture and the Environment: Perspectives on Sustainable Rural Development, Ernst Lutz, ed. Washington DC: The World Bank, 1998
2. Barrig, M.1991. "Women and Development in Peru: Old Models, New actors" in Environment and Urbanization, 3(2) October.
3. Beall, J. 1996. "Participation in the City: Where Do Women Fit In?", in *Gender and Development*, 4(1) February.
4. Chant, S. 1996. Gender, Urban Development and Housing. Habitat Monograph Series #3. UNDP. New York.
5. Falu, A. and M. Curutchet. 1991. "Rehousing the Urban Poor: Looking at Women First" in *Environment and Urbanization*, 3(2) October.
6. Hamerschlag, Kari and Annemarie Reerink, 1996. *Best Practices for*

Gender Integration in Organizations and Programs from the Inter Action Community: Findings from a Survey of Member Agencies. Washington, D.C.: Inter Action.

7. Machado, L.1987. "The Problem for Women-headed Households in a Low Income Housing Programme in Brazil", in Moser, C. and L. Peake. eds. *Women, Housing and Human Settlements.* Tavistock Publications.
8. OECD. 1995. Women in the City: Housing, Services and the Urban Environment. Paris: OECD.
9. Rai, S. 1995. "Women and Public Power: Women in the Indian Parliament" in *IDS Bulletin*, 26(3).
10. Robinson, M. 1995. "Introduction, Towards Democratic Governance", in *IDS Bulletin*, 26(2) April.
11. Tinker, I. and Jaquette, J., 1987: UN Decade for Women: Its Impact and Legacy. *World Development*, 15 (3): 419-427.
12. Yudelman, S. W., 1987: The Integration of Women into Development Projects: Observations on the NGO Experience in General and in Latin America in Particular. *World Development*, 15, Supplement: 179-187.
13. UNCED (United Nations Conference on Environment and Development), 1992: Agenda 21. Rio de Janeiro.

18

A Spatio-Temporal Analysis of Flood Disasters in North Bihar Plain: Socio-Economic and Environmental Impacts and Mitigation Strategies

G.S. Chauhan

Introduction

Floods are considered to be the most devastating hazards among all the natural calamities. Floods are a global phenomenon and severe floods frequently occur almost every year in various isolating regions of the world causing immense loss of life, large scale damage to property and untold miseries to millions of people. Hence, it may be rightly stated that such types of natural calamities left behind a story of death, hunger, epidemic and mass destruction. In case of India, it has been observed that most parts of northern India are more frequently hit by the severe floods because the entire region is ecologically fragile and dominated by the flood prone rivers system. Though, such hazards also occur in the various other isolated parts of the country including low lying coastal areas relatively with moderate frequency. It is a fact that floods have an intricate relationship with environment degradation which could be relatively called manmade disaster rather than natural disaster. Hence, flood hazards are precisely called natural since they result from a set of natural phenomena connected directly with the atmosphere and surviving topographical structure.

Such hazards undoubtedly cause sudden disruption to the normal life and immense damage to the biotic and a biotic activities to such an extent that social, economic and bio-physical mechanisms available to the living populous become inadequate to restore normalcy. Similarly, such vagaries of nature have a far reaching adverse impact not only on the human and livestock population but also in totality on various anthropogenic activities posing a severe threat to the fragile eco-system, socio-economic development and ecological sustainability of the affected region. Moreover, such hazards also exert a heavy toll of damage on the flora and fauna and ultimately destabilize the national and regional economy.

Objectives

- To demarcate and identify the flood-prone areas.
- Spatio-temporal analysis of level of flood damages.
- To examine and to assess the estimation of flood loss.
- To highlight the major causal factors of flood disasters.
- Meaningful and relevant strategies to mitigate flood disasters.

Material and Methodology

The research study is primarily based on the secondary data sources only. The major data and information pertaining to the research study have been collected from the various secondary sources only. The main sources of secondary data are Disaster Management Department including Dept. of Floods and Irrigation, Government of Bihar. Apart from that extensive flood literature have been collected from Central Water Commission, Ministry of Agriculture and from the National Disaster Management Authority, Government of India. Similarly, various latest flood reports related to flood management and water resource department, Government of Bihar have also been consulted to get the latest information regarding the recurring floods of Bihar including the study area. Precisely, all the data has been analysed by applying various statistical methods and techniques, for instance, circular diagrams, comparative bar diagrams and line graphs have been used including some maps and tables have also been prepared to depict the spatio-temporal view of North Bihar flood disasters. Likewise various flood areas, figures and data have been analysed and highlighted with the help of maps, charts and tables including all other related relevant information to understand the extent and magnitude of the existing problem of floods in North Bihar plain.

Results and Discussion

Global Flood Scenario

According to an estimate about 3.5 per cent of the total geographical areas of the globe are covered by flood plains, which house about 16.5 per cent of the total population of the world. The most notorious rivers of the world in terms of devastating floods and resultant damage to the natural ecosystem, loss of human lives and property are the Ganga and its tributaries such as the Yamuna, the Ramganga, the Gomati, the Ghaghra, the Kosi, the Damodar, and the Brahamaputra followed by the Mahanadi, the Krishna and the Godavari. The other famous and important rivers which are liable to bring furious and disastrous floods are the Yellow river (Huang He) and the Yangtze river of China, the Yellow, the Irrawadi, the Indus, the Niger, the PO, the Euphrates and the Tigris. The Mississippi and Missouri rivers of the USA are the most notorious rivers of the world with regard to havocs caused by them at regular intervals. The statistics of last more than 100 years or so reveal that flood disasters occur every 6 to 10 years in the Yangtze delta area of China. The major and deadliest flood disasters of the world in terms of human lives loss during 1931 to 2008 are the Chinese floods of 1931, 1935, 1938, 1939, 1948, 1954, 1975, 1981, 1989, 1991, 1996, 1998, 2002, 2005, 2008 followed by India's floods of 1943, 1955, 1961, 1968, 1978, 1979, 1992, 1993, 1998, 2008, 2005. The world flood disaster years of Bangladesh were 1974, 1987, 1988, 1993, 1998, 2004. Similarly various other countries of the world have also experienced the heavy floods in their region. For instance, Iran in 1954, Peru in 1941, 1962, Afghanistan in 1992, Pakistan in 1950, 1977, 1992, Japan in 1953, 1957, Algeria in 1927, Italy in 1963, Brazil in 1966, 1967, 1970, 1974, 2004, USA in 1913, 1928, 1937, 1951, 1955, 1993, 1977, 2005, Indonesia in 2003, U.K. in 1966, 2007, Russia in 2002.

Flood Scenario in India

With it vast territory, large population and unique geo-climatic conditions, the Indian subcontinent is exposed to natural catastrophes traditionally. The natural hazards like floods are not rare in the Indian arena. The Indian subcontinent is amongst the world's most disaster-prone area with nearly 68 per cent, out of which only 5 per cent area is considered to be vulnerable to floods. India being one of the richest countries with regard to its water resources, is continuously suffering from the menace of floods for centuries, due to the poor management of this rich natural resource and unplanned development has been another cause of floods. Because of the large geographical area (32,87,263 sq. km.)India often faces severe flood hazards occurring frequently in its different parts throughout its history of civilization. The country receives an annual precipitation of 400 million ha metres of the annual rainfall, 75 per cent receives four months of monsoon (June to September) and as a result almost all the rivers carry a heavy load of discharge during this period. The problems of sediment deposition, drainage congestion and synchronization of the river beds compound the flood hazard in various parts of the country.

It has been estimated that over 90 per cent of the total damage to the property and crops in India is done in the plains of North India. The total area affected annually on an average is 3.5 million ha. and it was as high as 10 m ha in 1988, the worst year. The average annual damage to crops, house and public utilities during the period 1953-96 was about Rs. 46303 million in the floods of 1988. As it has been already mentioned that India predominantly the state of Bihar time immemorial has been frequently hit by floods. There is hardly a year when some or other parts of the country does not face the spectre of floods. With regard to recurring floods the trends are quite significant. In spite of the state and central government flood policy and several flood control schemes, the loss and damage due to floods clearly appears to be increasing. The large part of population being subjected to distress is rising in flood prone areas. The sharp ups and downs in the flood trend seems to be occurring after 1965, with the worst year of 1978, 1985 followed by 1988. Almost parts of northern India particular Indo-Gangetic and Brahmaputra plain area are intensively affected by severe floods with moderate to high intensity. But out of them, Bihar, U.P., Assam, including West Bengal are considered to be the worst affected states. The other relatively equally affected states are Punjab, Haryana, Himachal Pradesh followed by Andhra Pradesh, Karnataka, Kerala, Tamil Nadu, Maharashtra, Gujarat and Rajasthan.

Flood Scenario in Bihar State

Bihar is probably the most important state, worst affected by the severe floods almost every year. Millions of people have been subjected to misery from time to time due to the cruel vagaries of the Kosi in North Bihar. On an average, nearly 27 per cent of flood damages in India are accounted for in Bihar. Annual destruction of property and crops by the Kosi amounts to Rs. 100 million. Being a very destructive river, the Kosi is said to be "India's River of Sorrow". Actually, the entire stretch from Vaishali-Muzaffarpur to Darbhanga is the major flood-prone area in the state. In this belt Kharif crops are the major casualty and vast stretches of paddy fields appear sometimes like a sea of muddy water. Moreover, the Kosi sometimes creates an unpleasant scenario in which it is very difficult to differentiate the river from flood water. The districts which have been identified as worst affected are Dharbhanga, Madhubani, Sitamarhi and Saharsha. The other moderately affected districts are Muzzarfur, Begusarai, katihar, Bhagalpur, Mungher and Purnia including Godda.

The Study Region

The North Bihar Plain is that part of the middle Ganga plain which is located in Bihar state. The region is very fertile with a moderate temperature and adequate rainfall. It extends from 250 15' to 270 to 31' North latitude and from 830 22' to 880 20' East longitude. The total geographical area of the study region is approximately 52,385 sq. kms. comprising districts namely Muzaffarpur, East Champaran, Sitamarhi, Seohar, Saharsa, Supaul, Darbhanga, Madhubani, Khagaria, Samastipur, Begusarai, Araria, Madhepura, Purnea, Katihar, Kishanganj, Saran, Gopalganj. West Champaran, Vaishali and Siwan. The region is bounded by Nepal in the north, West Bengal in the east, river Ganga in the south and Uttar Pradesh in the west.

Spatio-Temporal Extent of Floods of North Bihar Plain

The study region is the most flood-prone region in the entire country which accounts for about 17 per cent of the total flood affected area of India. Out of 22 districts of the study region, 19 districts have been identified as flood affected. Almost all districts including their concerned villages were affected and influenced (Fig. 1)

Fig. 1 : *Map Showing Flood Zones in Bihar State*

The study region is the most flood-prone area in the entire country which accounts for about 17 per cent of the total flood affected area of India. Out of 22 districts of Bihar State, 19 districts of the study region have been identified as flood-prone districts (Table 3) As it has already been mentioned that the study region has already confronted several disastrous and furious floods during the period of the last 29 years or so (1979-2007) with different levels of damages. The least affected area was recorded merely 0.76 lakh ha. during 1992 against 27 lakh ha. highest affected area during 2004. The similar trends have also been observed in case of cropped area affected by floods. In the same year (1992) only 0.25 lakh ha. cropped area was hit by flood disaster, the lowest figures during the given time period which has further touched the highest figure that is about 16.42 lakh ha. cropped areas in 2007 (Table 2). The study region has experienced flood disasters of different extent and magnitude. But the recent fierce floods of 2007 proved very disastrous and distinctive in the state of Bihar particularly in north Bihar plain's economy as well as its agricultural crops. It has been estimated that about Rs. 370981.07 lakh (Rs. 3709.81 million) property including crops, houses and public property was hit and damaged by these floods. About 12.52 lakh ha. cropped area was damaged of the value of Rs. 130508.22 (Rs. 1305.08 million). Ultimately, the existing situation was very alarming and critical. As it has been already stated that out of 22 districts of Bihar 19 districts of north Bihar plain have been identified as flood-prone during the flood of 2007. About 10830 villages study region were in the grip of flood fury and the **worst affected districts** were (i) Darbhanga (2104 villages), (ii) Muzaffarpur (1704 villages), (iii) Vashali (1306 villages) and (iv) East Champaran (1159 villages). **Moderate flood affected districts** were (i) Samastipur (842 villages) (ii) Madhubani (836 villages), (iii) Sitamarhi (806 villages), (iv) West Champaran (493 villages), (v) Begusarai (346 villages), (vi) Katihar (319 villages) and (vii) Khagaria (208 villages). Whereas the **least affected districts** were (i) Sharsa (184 villages), (ii) Sheohar (150 villages), (iii) Purnia (107 villages), (iv) Gopalganj (104 villages), (v) Supaul (194 villages), (vi) Madhepura (46 villages), (vii) Arariya (25 villages), and (viii)

Siwan (2 villages) (Table 3). The same kind of trends have also been observed in case of the damage cropped area by these districts (Fig. 2 and 3)

Main Causal Factors of North Bihar Flood Plain

1. Shifting courses of the Kosi and sand and silt deposition along its courses.
2. Uncontrolled and indiscriminate development of flood plain areas due to pressure of accelerating pressure of population.
3. Heavy and excessive rainfall in river catchment.
4. Meandering courses of existing rivers and their tributaries.
5. Large scale deforestation in the catchments and further no afforestation in the area.
6. Increasing level urbanization in the flood-prone areas.
7. Slopes of river system.
8. Blockade of natural flow of the river system.

Impact of Flood Hazards

In fact, the disasters are the complex and stern events of nature which pose a severe menace to the various types of anthropogenic activities of the society and their impact on human life is multi-dimensional. Almost all aspects of social life, i.e. domestic, social, economic and cultural are severely affected by such types of hazards. However, it is very difficult to assess the indirect environmental damage since plant - animal - human chain is intensively affected by such calamities. The direct impact of flood hazards is indeed tremendous especially on the economy. The damage done by such hazards to infrastructure, crops, and productive assets of local population is massive especially on its poor strata besides a huge financial burden of relief and rescue operations.

Indirectly these often lead to decline in production, loss of income, unemployment, indebtedness of the poor and increased cost of goods and services. Actually, the large scale damage caused by floods run into thousand million rupees. But worst of all are the precious lives of human-beings including their cattle wealth, lost in hundreds and thousands in case of cattle. What is most surprising is that despite thousand millions of rupees spent on flood control measures every year, the losses keep mounting. Although, the figures may fluctuate from year to year. However, floods that occurred in the past also damaged the life and property of the society. But their overall impact was not felt in the past because of low population pressure, and extensive industrial following other development activities in the plain areas, as is the case presently. Moreover, such hazards also exert a severe threat to the existing fragile eco-system and environmental status. Apart from this countless trees especially on the roads, canals, river and railway line sides uprooted conditions prevailed in the most affected regions.

Affected Area

The statistical figures of the given time period (1979-2007) reveal that there are many ups and downs with regard to area prone to flood hazards in North Bihar Plain. The area affected by floods was 8.06 lakh ha in 1979 against 1.81 lakh ha in 2006 with a peak of 47.50 lakh ha during 1987 and an average of 13.76 lakh ha affected annually during the period of 29 years, i.e. 1979-2007. Surprisingly, the total area affected has varied only between 17.86 lakh ha in 1980 to about 27 lakh ha on average during the last few years; the only exception was 1978 when the total area affected exceeded 47 lakh ha (**Table 1**).

Affected Population (Human and Livestock)

In addition, there is a great loss of human lives and livestock often affecting the poor part of the rural population. So it is a true fact that hazards like floods have indeed a very intensive impact on human and livestock that is the only wealth of rural people. The affected population particularly in the rural arena faces a number of problems but out of them, food, shelter and clothing are affected severely. The average affected population during the designated time period has been recorded has been recorded 77.23 lakh million, with maximum 286.62 lakh million (**Table 1**) in 1987. The tentative affected population according to the latest figures of 2007 was registered 244.46 lakh relatively higher than the average. The loss of economic wealth and property caused by severe floods often run into thousands of millions rupees. But all of them, the worst strike is the precious and priceless lives of hundreds lost in such vagaries of nature. Hence, the concept of casualty of human beings is totally governed by the degree and extent of hazards magnitude in the flood-prone areas. The average loss of human lives was more than 235, with highest 1399 in 1987 alone. While the cattle loss was relatively higher compared to human loss in the same duration. The average cattle loss was relatively higher compared to human loss in the same duration. The average cattle loss was noticed to be 741, whereas the worst-cattle strike year was 1987 in which nearly 5302 cattle lost their lives (Table 1). Moreover, occasionally hundreds of livestock fled and as many as died in the wake of fodder resources in the flood-prone areas. On an average, the natural disasters take a toll of over 235 human lives and damage 0.22 million houses annually. Past statistics reveal more than 50 per cent human lives were lost due to collapse of houses and buildings.

Damaged Cropped Areas

Every year thousand ha land including fertile and productive cropped land is severely damaged both in the hilly areas as well as in the plain areas of the study region. As a result there was a huge loss of nutrients and other supporting elements required for genetic growth of crops and plant. The total cropped area affected annually is about 7.03 lakh ha. and was as high as 25.70 lakh ha in 1987, the worst year in the past 29 years (Fig. 4&5).

The agricultural land dominated by the kharif cropped area is probably the first victim of floods in the vast plain areas of North Bihar plain. Hence, the flood hazards have a great impact on the rural agriculture system which is the base of local economy, consequently thousand hectares of crop land were either converted into barren land or attained the status of degradation. In case of crops, they are badly hit and damaged and ultimately costing Rs. 15,807 million in the year 2007. The total loss on account of flood damage to crops, houses, and public utilities was estimated at Rs. 3,869 million and was as high as Rs. 38,409 million in 2007,

(Fig. 2) the worst year in the past 29 year in the country. Total damage caused by floods is estimated to the tune of Rs. 38409 million during 2007 at an average of Rs.3869 million between 1979-2007. The yearly flood loss during the year 1979-2007 has been shown in the (Table 2).

Precisely, the flood disasters have the following impacts on the various anthropogenic activities including two major sectors, i.e. agriculture and socio-economic conditions of society:

I. Every year millions of people become homeless, rendered for shelter for many days and most of the cases are forced to stay under the open sky.

II. Thousands to millions of houses and settlements have been badly damaged and a large number of them collapsed.

III. Similarly thousands to million hectares of agricultural land come under deep flood water and not in a condition for further cultivation.

IV. Millions tons of fertile top soil have been eroded by several major rivers and their tributaries of the country and ultimately deposited in the seas/oceans.

V. Hundreds of people fled in the flood water and an equal numbers have died either due to lack of food availability or epidemics.

VI. Thousands of hectares of land have been converted into waste land/barren land and resultantly problems of salinity and alkalinity including water logging originate.

VII. Due to standing of a large quantity of flood water at certain places for a long time, various types of water-borne diseases and ground water table suddenly rose.

VIII. Thousands of livestock either fled in flood water or died in the wake of fodder shortage.

IX. Due to overflow of flood water in various rivers, tributaries, canals and drains there always remained the threat of breaches and seepage at several vulnerable points.

X. Production of certain agricultural crops including cash crops either drastically declined or lost their quality and quantity.

XI. National and state highways including their other associated link roads have been submerged in flood water subsequently failure of traffic for several weeks or so resultantly heavy disruption of economic and commercial activities.

Table 1: Flood Affected Area and Flood Damages in Bihar (Abstract for the Period 1979-2007)

Sl. No.	*Item*	*Unit*	*Average Flood Damage*	*Maximum Damage With Year*	*Damage During 2007*
1	Area Affected	Lakh ha.	13.76	47.50 (1987)	N/A
2	Population Affected	Lakh	77.23	286.62 (1987)	244.46
3	Human Lives Lost	Nos.	235.24	1399 (1987)	948
4	Cattle Lives Lost	Nos.	741.93	5302 (1987)	988
5	Cropped Area Affected	Lakh ha.	7.03	25.70 (1987)	16.42
6	Value of Damage to Crops	Rs. Millions	1833.31	13320.40 (2007)	13320.40
7	House Damages	Millions	0.22	1.70 (1987)	0.69
8	Value of Damage to Houses	Rs. Millions	1131.09	9281.59 (2007)	9281.59
9	Value of Damage to Public Utilities	Rs. Millions	1010.06	15807.25 (2007)	15807.25
10	Value of Damage to Houses, Crops and Public Utilities	Rs. Millions	3869.97	38409.24 (2007)	38409.24

Source: Disaster Management Department, Government of Bihar, October 11, 2007.

Table 2: Flood Damages in Bihar (1979-2007)

S. No.	Year	Area Affected (Lakh hect.)	Number of Affected				Number of Lives Lost		Damage of Crop		Damage to Houses		Value of Public Property	Value of Damage to Crops, Houses and Public Property
			District	Village	Human	Cattle	Human	Cattle	Area (Lakh hect.)	Value (Rs. Lakh)	Number	Value	Damaged (Rs. Lakh)	(Rs. Lakh)
1	1979	8.06	13	N/A	37.38	-	14	4	2.74	1901.52	27816	103.4	-	2004.88
2	1980	17.86	21	7010	74.45	-	67	42	9.43	7608.43	118507	561.3		8169.74
3	1981	12.61	21	7367	69.47	74.83	18	11	7.71	7213.19	75776	406.7	-	7619.88
4	1982	9.32	15	3708	46.81	45.14	25	14	3.23	9700.00	68242	686.5	955.33	11341.85
5	1983	18.13	22	4060	42.41	-	35	21	5.78	2629.25	38679	172.4	258.14	3059.83
6	1984	30.5	23	11154	13.50	-	143	90	15.87	18543.85	310405	2291.54	2717.72	23553.11
7	1985	7.94	20	5315	53.09	-	83	20	4.38	3129.52	103279	756.20	204.64	4090.38
8	1986	19.18	23	6509	75.80	-	134	511	7.97	10513.51	136774	647.20	3201.99	14362.74
9	1987	47.5	30	24518	286.62	33.25	1399	5302	25.7	67881.00	1704999	25789.32	680.86	94351.18
10	1988	10.52	23	5687	62.34	0.21	52	29	3.95	4986.32	14759	211.32	150.64	5348.28
11	1989	4.71	16	1821	18.79	0.35	26	-	1.65	709.88	7746	160.73	83.70	949.31
12	1990	8.73	24	4178	39.57	2.70	36	76	3.21	1818.88	11009	160.12	182.27	2161.27
13	1991	9.80	24	4096	48.23	5.13	56	84	4.05	2361.03	27324	613.79	139.93	3114.75
14	1992	0.76	8	414	5.56	0.75	4	-	0.25	58.09	1281	16.14	0.75	74.98
15	1993	15.64	18	3422	53.52	6.68	105	420	11.35	13950.17	219826	8814.00	3040.86	25805.03
16	1994	6.32	21	2755	40.12	15.03	91	35	3.50	5616.33	33876	494.77	151.66	6262.76

17	1995	9.26	26	8233	66.29	8.15	291	3742	4.24	9514.32	297765	75100.44	2183.57	29208.33
18	1996	11.89	29	6417	67.33	6.60	222	171	7.34	7169.29	116194	1495.34	1035.70	9700.33
19	1997	14.71	26	7043	69.65	10.11	163	151	6.55	5737.66	174379	3056.67	2038.09	10832.42
20	1998	25.12	28	8347	134.70	30.93	381	187	12.84	36696.68	199611	5503.70	9284.04	51484.42
21	1999	8.45	24	5057	65.66	13.58	243	136	3.04	24203.88	91813	5384.95	5409.99	34998.82
22	2000	8.05	33	12351	90.18	8.09	336	2568	4.43	8303.70	343091	20933.82	3780.66	33018.18
23	2001	11.95	22	6405	90.91	11.70	231	565	6.50	26721.79	222074	17358.44	18353.78	62434.01
24	2002	19.69	25	8318	160.18	52.51	489	1450	9.40	51149.61	419014	52621051	40892.19	144663.31
25	2003	15.08	24	5077	76.02	11.96	251	108	6.10	6266.13	45262	2032.10	1035.16	9333.39
26	2004	27.00	20	9346	212.99	86.86	885	3272	13.99	52205.64	929773	75800.51	0.00	128015.15
27	2005	4.60	12	1464	21.04	5.35	58	4	1.35	1164.50	5538	382.79	305.00	1852.29
28	2006	1.81	14	959	10.89	0.10	36	31	0.87	706.63	18637	1225.03	8456.17	10387.83
29	2007	N/A	22	11850	244.46		948	988	16.42	133203.97	690237	92815.88	158072.46	384092.39

Source: Disaster Management Department, Government of Bihar, October 11, 2007

Table 3: Flood Damages in North Bihar Plain, 2007

S. No.	District	Number of Affected			Population	Number of Line Lost		Damage to Crops		Damage to Houses		Value of Public Property Damage (Rs. Lakh)	Value of Total Damage to Crops Houses and Public Property (Rs L.)
		Block	Pan-chayats	Village	Affected (Lakh)	Hu-man	Cattle	Area (Lakh Hect.)	Value (Rs. Lakh)	Number	Value (Rs. Lakh)		
1	Darbhanga	18	322	2104	34.41	136	395	1.75	6606.1	83127	13106.14	18271.02	37983.26
2	Muzaffarpur	15	366	1704	32.63	91	190	1.24	12663	6550	11073.00	24951.00	48687
3	East Champaran	27	385	1159	30.96	96	28	1.58	15400	52840	8278.31	-	23678.31
4	Sitamarhi	17	278	806	27.86	33	104	0.51	7803.94	103193	16084.85	63618.25	875007
5	Samastipur	19	381	842	21.13	175	75	1.25	16710.07	29391	775.00	17896.46	35381.53
6	Vashali	15	266	1305	18.75	32	11	0.73	6582.47	26012	2501.14	2977.35	12060.96
7	Madhubani	20	331	835	18.07	63	29	1.39	7936.23	96322	9116.45	25733.68	42786.36
8	Khagaria	7	108	203	10.10	108	71	0.50	8507.33	32500	1945.08	372.00	10824.41
9	Begusaria	10	72	346	7.59	48	15	0.94	16057.84	40740	11773.10	444.00	28274.94
10	West Champaran	16	174	493	7.08	20	7	1.03	21117.99	12861	564.60	244.32	21926.91
11	Katihar	16	86	319	5.40	36	1	0.40	2768.85	2299	216.62	33.65	3019.12
12	Sheohar	5	46	150	3.86	1	-	0.25	693	50728	6477.10	105.00	7275.1
13	Sharsa	6	68	184	3.65	25	-	0.25	861.4	16383	935.25	140.38	1937.03
14	Supaul	6	35	94	2.54	1	-	0.25	574.84	15000	300	17.75	892.59
15	Purnea	4	27	107	1.70	-	-	-	-	4783	-	-	-
16	Gopalganj	8	44	104	1.64	10	1	0.40	6109.16	5500	321.85	2157.5	8588.51
17	Madhepura	3	24	46	0.70	19	-	0.05	116	2100	32.00	10.00	158
18	Arariya	1	12	25	0.04	-	-	-	-	12	-	-	-

19	Siwan	1	1	2	0.02	-	-	-	-	-	-	-	-
	North Bihar Plain	**214**	**3026**	**10830**	**228.13**	**894**	**927**	**12.52**	**130508.22**	**639341**	**83500.49**	**156972.36**	**370981.07**
20	Bhagalpur	13	161	340	6.10	37	-	1.51	900	24321	2445.00	1000.00	4345
21	Nalanda	18	150	553	6.03	17	61	1.89	1746.01	15582	6869.22	100.10	8715.33
22	Patna	6	28	127	4.20	-	-	0.50	49.74	10993	1.25	-	50.99
	South Bihar Plain	37	339	1020	16.33	54	61	3.90	2695.75	50896	9315.47	1100.10	13111.32
	Bihar State	251	3365	11850	244.46	948	988	16.42	1335203.47	690237	92815.96	158072.46	384092.39

Source: Disaster Management Department, Government of Bihar, October 11, 2007

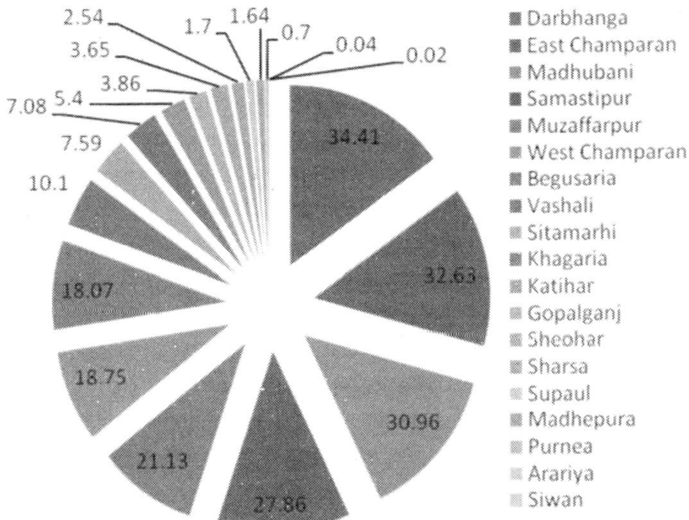

Fig. 2: *North Bihar Plain Flood Affected Area During 2007 (In Lakh ha.)*

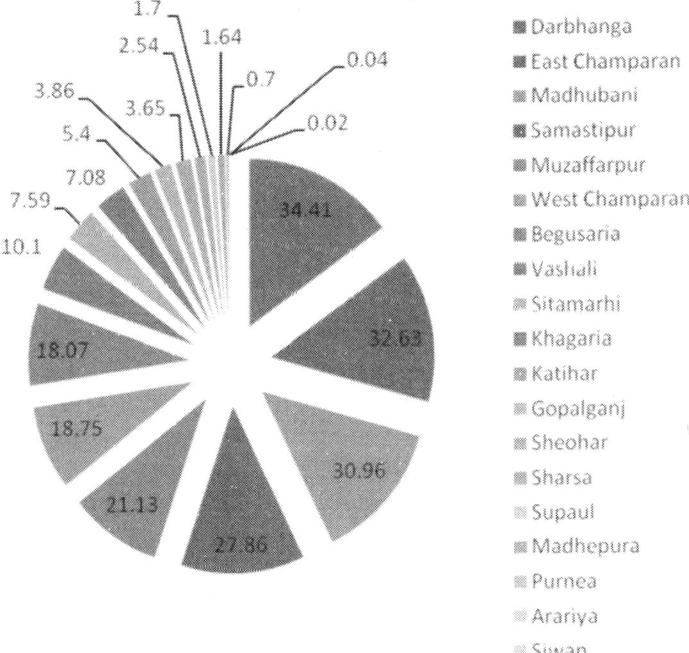

Fig. 3: *North Bihar Plain Population Affected Area During 2007 (In Lakh)*

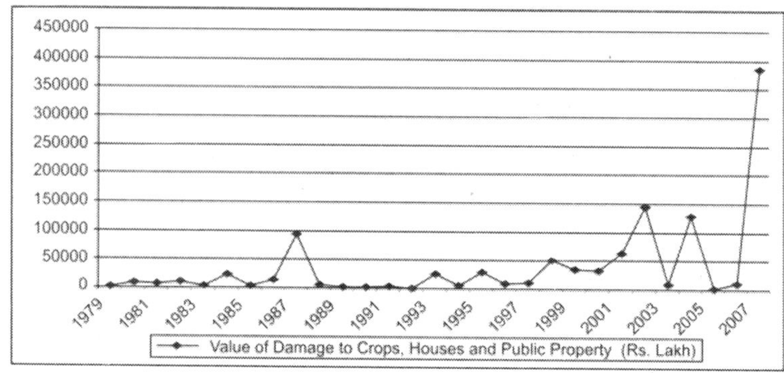

Fig. 4 : *Annual Flood Damages in Bihar State (1979-2007)*

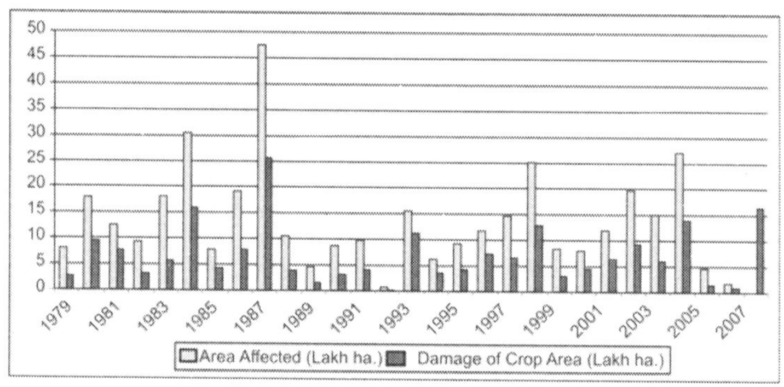

Fig. 5 : *Flood Damages in Bihar (1979-2007)*

Preventive Measures for Floods Mitigation

It is a well accepted fact that occurrence of floods cannot be prevented. Though, their adverse societal and economic impact can be reduced substantially by undertaking various preparedness and mitigation measures by active community involvement. Minimizing the loss of precious human life is the first priority in flood management. Almost each and every year stories of misery, devastation, death and epidemics are repeated that monsoon leaves in its aftermath. But with poor planning, mismanagement of water resources, lack of political will and meagre money, it is a very difficult task to control the frequent floods.

Hence, it is necessary to suitably "*Manage Floods*" with a view to reduce the damage potential and avoid loss to lives of humans and cattle. So, in this direction "*A better understanding of behaviour of rivers*" could help in order to prevent loss due to flood hazards. The Central as well as State Governments from time to time review such measures and take appropriate and meaningful steps but even then, many things have gone wrong with most of the damage in floods being due to increase in population along the banks. Another is mass destruction of forests to earn a livelihood, reclamation of more and more lands even within the riverside areas have caused changes in the river regime system over the years. All these have led to increasing flood damage to various control measures undertaken in the country as a whole and the study region in particular.

In the past the general approach to tackle the problem of floods was to construct productive bunds or drainage channels for their houses and even cultivated lands. They were constructed, however only when there was immediate damage of floods. This approach had its ancient origin and tradition in the study region and various parts of India, because flood protection embankments have been extensively constructed in the areas of Indo-Gangetic plain. But in the modern period many things have transformed and numbers of new techniques and methods have been used by the flood experts to reduce and mitigate flood loss.

Strategy for Flood Management and Sustainable Development

In order to achieve sustainable development, the model Flood Plain Zoning Bill should be introduced to all the Indian states including the state of Bihar. Although, the bill came into force in 1975 but only Manipur enacted legislation on the basis of the model bill, keeping in mind the relevance and importance, the adoption of the bill should be made necessary so that fruitful results may come forward in the affected region. By human settlement in the flood plain of a river. To achieve the same goal, massive afforestation programmes should be launched along the river, canals and drains to strengthen their embankments. At local levels, the drainage system should be made wider, deepened and strengthened generally in those areas which are frequently flood-prone. On a large scale drive "*clean up*" should also be started in all the vulnerable rivers and their tributaries including canals in the presence of expert engineers right before the occurrence of the rainy season so that deposited silts, sand and debris be removed so that is may be helpful to increase the carrying capacity of rivers and canals.

Moreover, another but most suitable and useful method for mitigation of floods is the flood plain zoning. It is considered the most effective and reliable method for flood prevention but before identification for flood plain zoning, there is an urgent need for the better understanding of behaviour of rivers, that can help to prevent occurrence of floods. So, flood-prone areas should be identified and mapped. Aerial surveys should be made especially to assess loss of agricultural crops, vegetation cover and property to the exact flow of flood water. During critical flood hours, along the sides of river and canals strict vigil by army/police personnel should be taken in order to face any eventuality and to prevent local population. Apart from all these concerted efforts and techniques described in the strategy, the other one more significant and important method is the advance warning system which is now easily possible and easier with the help of satellites and INSATS. By adopting all the stated methods, techniques including strategies, it may be possible to achieve the sustainable development and can save the millions and avoid the huge burden of the national economy spent on flood measures. Moreover, it will also be helpful to maintain the ecological balance of the affected areas.

Conclusions

It has been observed that most parts of northern India are severely affected by flood hazards particularly Indo-Gangetic-Brahmaputra plain. Almost all the anthropogenic and commercial activities are badly hampered on a large scale by the flood hazards during the past 44 years subsequently destabilization of the national economy. So, in order to mitigate flood hazards, there is an urgent need first of all to identify and map flood-prone areas as suggested in the above discussions. Secondly, the new modern techniques should be applied for advance warning system which is possible now through various satellite and remote sensing services. Hence, the flood forecasting and warning system should be adopted because it is one of the most reliable and cost effective methods and moreover, over the years there is considerable improvement in the methodology and acquisition of the latest technology. Dams, embankments and reservoirs should be launched involving local people especially women and school children along the rivers, streams, canals including drains to strengthen their embankments and similarly special clean up drives should also be launched in order to remove the deposited debris and silt in all the vulnerable rivers, canals and drains (at local level) in the presence of expert engineer right before the occurrence of the monsoon season. By adopting all these said measures and flood management works in the country, it would be possible to save the precious lives of human beings including cattle wealth and considerably reduce the immense flood damages in the country.

REFERENCES

Aggarwal, A. Ed. (1991). *Floods, Plains and Environmental Myths*, CSE, New Delhi.

Ansari, A.A. (1987). *The Kosi: A Study in River Regime*, Unpublished Ph.D. Thesis, Dept. of Geography, University of Delhi.

Beven, K. and Carling, P. (1989). *Hydrological, Sedimentological and Geomorphologic Implications*, John Wiley, New York.

Barmmer, H. (1968). "Flood in Bangladesh". *The Geographical Journal*, Vol. 156, pp. 12-22.

Chauhan, G.S. (2002). A Spatio-Temporal View of Indian Floods: Their Impact and Strategies for Their Mitigation, published in *Flood Defence* 2002, edited by Zhao-Y in Wang Science Press New York Ltd., Beijing.

Chauhan, G.S. (2004). Flood Hazards in India: Their Impact and Management for Sustainable Development, published in *Water Resource Management* eds., G.S. Chauhan and R.N. Dubey, Shree Natraj Prakashan, New Delhi, pp. 104-130.

Chauhan, G.S. (2004). Impact of 1995 Haryana's Flood: Mitigation

Measures and Strategy for Sustainable Development, published in *Water Resource Management* eds., G.S. Chauhan and R.N. Dubey, Shree Natraj Prakashan, New Delhi, pp. 131-152.

Chauhan, G.S. (2005). Flood Disasters in North Bihar Plain, published in *Sustainable Water Management*, Arvind Kumar, (ed.), Dept. of Botany, Dr. S.P. Mukherjee Government Degree College, Phaphamau, pp. 201-212, Allahabad.

Das, K.N. (1968). "Westward Shift in the Course of the Kosi". *NGJI*, Vol. XIV, Part I.

Desai, C.G. (1982). *The Kosi River: Its Morphology and Mechanics in Retrospect and Prospect*, C.W.C. New Delhi.

Flood Report (2007). Flood Management Information System Cell, First Annual Report, Water Resources Department, Government of Bihar, Patna

Flood Report (2009). Flood Management Information System Cell, Third Annual Report, Water Resources Department, Government of Bihar, Patna

Ganga Flood Control Commission (1993). *A Comprehensive Plan of Flood Control for the Kosi Sub-basin*, Report, Vol. I. Government of Bihar, Patna.

N.C.F. (1980). Report of the National Commission of Flood, Part I, New Delhi, pp. 67-120.

Singh, R.B. (1989). Monitoring Soil Loss and Forecasting Erosion Hazards in India Himalaya, *India Cartographer*, Vol. IX, I: 432-442.

Singh, R.B. (1992). Monitoring Soil Loss and Sediment Deposition in Indian Perspective, Erosion and Sediment Transport Monitoring. Programmes in River Basins, Jim Bogen (ed.) NWREA, Oslo: 138-142.

Singh, R.B. (ed.) (1999). *Disaster Management*, Rawat Publication, Jaipur, Delhi.

Science Reporter, September 1993, Vol. 30, No. 9, New Delhi.

Shelter: October 13, (1999). A HUDCO-HSMI-Publication Special Issue: Towards a Safer Millennium.

Contributors

A.K. Mudgal, Principal and Head, Department of Zoology, Government Degree College, Aron, Guna.

A.R. Khan, Professor and Head, Department of Chemistry, Government P.G. College, Sheopur.

Alok Kumar, Associate Professor, Department of Economics, St. John's College, Agra.

Archana Shrivastav, Principal and Head, Department of Life Sciences, College of Life Sciences, Cancer Hospital & Research Center, Cancer Hill, Lashkar, Gwalior.

Arvind Dohre, Assistant Professor, Department of Botany, Government P.G. College, Sheopur.

D.K. Sharma, Head, Department of Zoology, Government Girls College, Shivpuri.

G.S. Chauhan, Ph.D., Geography, University of Delhi; Education Officer, University Grant Commission, Central Regional Office, Bhopal.

Gaurav Jaiswal, Lecturer, Department of Management, Prestige Institute of Management, Gwalior.

K.S. Rathod, Professor, Department of Management, Vikrant College, Gwalior.

K.S. Sengar, Principal, Government P.G. College, Sheopur.

Kamakshi Dandotiya, M.B.A., Ph.D., Research Scholar, Morena.

Kaushal Gupta, Assistant Professor, Department of Political Science, Deshbandhu College, University of Delhi, Delhi.

Keshav Singh Gurjar, Associate Professor, Department of Physical Education, Jiwaji University, Gwalior.

Kumud Dandotiya, M.B.A., Research Scholar, Morena.

Manoj Sharma, Government P.G. College, Sheopur.

Mukesh Sharma, Assistant Professor, Department of English, Government K.R.G.P.G. College, Gwalior.

Neha Mathur, Assistant Professor, Department of Management, National College, Gwalior.

Niharika Sengar, Ph.D. Gwalior University; (R) 129/1, G Sector, Rajharsh Colony, Kolar Road, Bhopal.

P. Chaudhry, Assistant Professor, Department of Political Science, Government P.G. College, Sheopur.

Poonam Dubey, Ph.D., Scholar and Social Activist, Chaudhary Charan Singh University, Meerut.

Prashant Sahu, Assistant Professor, Department of Commerce, Maharaja Mansingh College, Gwalior.

Praveen Sahu, Head, Department of Commerce, Government Degree College, Mungawali, District Ashok Nagar.

R.K. Jain, Professor, Department of Biotechnology, Gandhi Vocational College, Guna.

R.S. Rathore, Assistant Professor, Government M.J.S. College, Bhind.

Ramendra Tiwari, Professor, Department of Botany, Dr. H.S. Gour University, Sagar.

Ramesh Kumar, Professor and Head, Government P.G. College, Sheopur.

Rashmi Sharma, Department of Chemistry, S.D. Government College, Beawar, Rajashthan.

S.D. Rathor, Professor and Head, Department of Commerce and Management, Government P.G. College, Sheopur.

S.K. Singh, Associate Professor, Institute of Management and Commerce, Jiwaji University, Gwalior.

S.N. Sharma, Head, Department of Hindi and Vocational Courses, Government P.G. College, Sheopur.

Sanjay Shrivastava, Assistant Professor, Department of Management, National College, Gwalior.

Sanjeev Gupta, Professor, Department of Commerce, Government M.L.B. College of Excellence, Gwalior.

Savita Shrivastav, Professor, Department of English, Government K.R.G. Autonomous College, Gwalior.

Shema Khan, Assistant Professor, Department of Chemistry, Government P.G. College, Dausa, Rajasthan.

Shobha Sharma, Associate Professor, Department of Physics, St. John's College, Agra.

Subhash Chand, Head, Department of Botany and Zoology, Government P.G. College, Sheopur.

Sudhir Pathak, Assistant Professor, Department of Botany, Government Girls College, Morena.

Sudhir Sharma, Assistant Professor, Department of Economics, Government M.L.B. College of Excellence, Gwalior.

Sunil Kumar, Commonwealth Fellow and Associate Professor, Department of Political Science, Shyam Lal College (Evening), University of Delhi, Delhi.

Sushil Gupta, Assistant Professor, Department of Chemistry, Government P.G. College, Sheopur.

Sweta Gupta, Lecturer, Preston College, Gwalior.

Tanu Middha, Assistant Professor, Department of Management, S.R.I.T.M., Banmore, Morena.